ORGANIC FOOD AND AGRICULTURE – NEW TRENDS AND DEVELOPMENTS IN THE SOCIAL SCIENCES

Edited by **Matthew Reed**

INTECHOPEN.COM

Organic Food and Agriculture – New Trends and Developments in the Social Sciences
Edited by Matthew Reed

CBS, Edition **2017**

Published by InTech
Janeza Trdine 9, 51000 Rijeka, Croatia

Publishing Process Manager Ivona Lovric

Technical Editor InTech DTP team

Cover Designer InTech Design Team

Image Copyright artjazz, 2011. DepositPhotos

Additional hard copies can be obtained from orders@intechopen.com

Organic Food and Agriculture –
New Trends and Developments in the Social Sciences, Edited by Matthew Reed
p. cm.
ISBN 978-953-307-764-2

INTECH

open science | open minds

Contents

Contents

Preface

The trajectory of organic food is never easy to predict, which makes it such a fascinating subject to study. It brings together controversies about science, society and nature onto the dinner plate, stressing out questions of global importance, such as 'what should I purchase and eat today?'. At present, the sales of organic food are going through a trough and the organic industry is consolidating as it learns how to operate in a new environment. The long boom in the key markets for organic products; North America, the European Union and Japan, is faltering and the domestic purchasing power of many people is increasingly constrained. Simultaneously organic agriculture, under the name of agro-ecology, is increasingly being presented as an answer to producing food sustainably, and improving the livelihoods of farmers in the global south. The recent report of the United Nations Special Rapporteur on the Right to Food, Olivier De Schutter, which recommends the global adoption of agro-ecology, is built on the sustained effort of academic researchers to demonstrate, through high quality research, the potential of organic agriculture (De Schutter 2011).

In the regions of the world where organic food is sold through markets, by which we mean the increasing chains of supermarkets or multiple-retailers, but also the farmers' markets, box schemes and subscriptions schemes, the concerns are different. Rather than those of technical execution or appropriateness for climatic conditions, they are more about how and why people chose to purchase certified organic products within a complex brandscape of competing claims on their attention and ultimately their purses. As many have argued, this focus on promoting organic food through the market place is not an accident but a deliberate strategy by the wider social movement surrounding organic food and farming. In a liberalized and globalized market for food, organic food has taken the challenge of the dominant model when it comes to consumers and the increasing amount of choices they have. Many other social movements have chosen to promote their cause through confrontation, lobbying or dramatic public protests, while the organic movement has opted to follow the route of radical consumerism.

This is a route that is not without controversy, as it attracts the carpetbaggers who are looking for a quick profit without sharing the ethics of the movement, and it also puts the movement's critique of how food is distributed into an ambivalent position. For some commentators it is not possible to contest the commodification of food whilst it is being stacked on the shelves of Wal Mart, Carrefour or Tesco. Yet, in the past two decades the entry of organic foods onto the shelves, frequented by the most powerful collective of consumers on the planet, has caused many changes. It has led to questioning of how food is produced, the risks taken by novel technologies in the food chain and ethics of agriculture, as well as the way in which the food market works in a globalised society.

The chapters and sections in this book reflect those discussions and how they have been developed within the social sciences. As one can imagine there is an emphasis on the way in which organic food is sold, the type of the stores, the operation of the market place, and how the decision making process is structured in the minds of consumers. There are also discussions on the food system and how it relates to the spatial practices of farm businesses, and the role of policy. We are fortunate to have authors from across the planet, in this book, who are attempting to understand how this global phenomenon has localised in their society. There are also lessons that can be learned, not only from places where organic food and farming is well established such as the US, Germany or the UK, but from where it is emerging and, as such, adapting to the aspirations of different societies. Until recently the literature about organic food and farming was limited because the dialogue has been a global one, and as the body of research has grown it is important that it remains so.

This book makes use of the recent appearance of the open access publishing. In contrast to the pattern of publications that dominate in academic community, where authors contribute their work for free and readers pay, in this book all the authors have paid to ensure that their work is freely available to readers. Whilst many academic discussions are available through books, frequently their limited publication runs mean that these works are rarely available as a paperback and are costly as hardbacks. Open access offers authors the opportunity to address a wider readership and perhaps to engage in a deeper dialogue than the more established routes of publishing. The fact that the authors have paid to cover the costs of making their work freely available does not mean that the thresholds of quality have been lowered; all the authors in this book are experienced in academic publishing and all of the papers have gone through a careful editorial process. The result is the book you are holding in your hands, or more likely, reading on the screen of your computer or e-reader. It represents a window into the scholarly discussion of organic food and agriculture. I would encourage you to take the opportunity to not only read the chapters but also engage with the authors and to foster a dialogue about the future of our food.

The organic movement started in the early twentieth century in response to the environmental threats that farmers, scientists, doctors and concerned citizens saw threatening global agriculture. At that time, they were worried about the decrease in the quality of food, the fact that planet Earth was turning more and more into a desert and it was getting harder to feed people. These are questions that are reoccurring in people's minds, and they remain relevant in today's society. (Reed 2010). As recent discussions of agricultural productivity have reminded us, since 1985 more than half of all the synthetic nitrogen fertiliser ever applied has been used; there are increasing concerns about the availability of phosphorous and the limits of an agriculture based on the use of fossil fuels (Cordell *et al.* 2009, Horlings & Marsden 2011). There is an increasing clamour for an intensification of agriculture that threatens to increase the technocratic control of agriculture, just as many social scientists are calling for a deepening of the democratic control of agriculture, and social movements are attempting to wrest sovereignty over their food, back from the global governance of the corporations and the world trade organisation. It is unlikely that there will be any lessening in the relevance of the arguments around organic food and farming, so the importance of rigorous enquiry by social scientists will remain of interest to, and in the interests of, a wide audience. This will require that social scientists are both open with their findings and methods but also open to the challenges that these times bring to their methods of social enquiry.

Dr. Matt Reed,
Countryside and Community Research Institute,
University of the West of England,
United Kingdom

Bibliography

Cordell, D., J.-O. Drangert and S. White (2009) The Story Of Phosphorus: Global Food Security And Food For Thought. *Global Environmental Change,* 19(2) pp. 292-305 ISSN 0959-3780

De Schutter, O. (2011) Report Submitted By The Special Rapporteur On The Right To Food, Olivier De Schutter. Pp. 21 in, (New York UN General Assembly Human Rights Council)

Horlings, L.G. and T. Marsden (2011) Towards The Real Green Revolution? Exploring The Conceptual Dimensions Of A New Ecological Modernisation Of Agriculture That Could 'Feed The World. *Global Environmental Change,* 21(2) pp. doi:10.1016/ j.gloenvcha.2011.01.004

Reed, M. (2010) Rebels For The Soil - The Rise Of The Global Organic Movement, (London: Earthscan) ISBN 1844075974

Part 1

Consumers and Markets

1

The Consumption Choice of Organics: Store Formats, Prices, and Quality Perception – A Case of Dairy Products in the United States

Ming-Feng Hsieh and Kyle W. Stiegert
University of Wisconsin-Madison
USA

1. Introduction

Consumers choose to purchase organic foods for a variety of reasons. Some of the commonly cited perceptions among consumers are that a) organic foods are grown without pesticides or other toxic chemicals and so they are healthier for them and their families, b) organic farming relies on more sustainable natural biological systems, which are better for the environment, c) practices and standards have evolved in the U.S. to improve the treatment of organically raised livestock. However, cropping and livestock systems used in organic farming tend to have higher costs per unit of output than in conventional farming. When these costs are successfully passed downstream, it ultimately means higher retail prices for those products that use the organic label. The price of organic food is typically 30-40%, and sometimes over 100%, more than conventional (non-organic) alternatives. The hefty price premium of organic food has been one of the major reasons for consumers to choose conventional over organic foods (Kavilanz, 2008). Wal-Mart in 2006 launched an aggressive "going green and organic" campaign that would greatly increase the number of organic products they offered with a price target of only 10% above the price for conventional counterparts. This market expansion and low pricing strategy has not only enhanced competition among food retailers in the United States but also encouraged consumers to rethink whether and where to buy organic foods: i.e. the choice of product type, organic or conventional, and the format of store in which the products were purchased.

Much research focuses on the rationales of how consumers make their store format choice (Bell et al., 1998; Bell & Lattin, 1998; Bhatnagar & Ratchford, 2004; Briesch et al., 2010; Ho et al., 1998; Hsieh, 2009; Hsieh & Stiegert, 2012; Messinger & Narasimhan, 1997; Tang et al., 2001). The studies show that the consumption in some product categories has stronger impacts on certain store format than the others when they are searching where to buy. In this research, we center our analysis on the other side of the question, that is, whether and how households patronizing different store formats would have different price sensitivities in making decision between organic and conventional alternatives for two dairy-case products. In particular, we examine the role of store format choice in households' consumption choice between organic and conventional alternatives for milk and eggs, two products that are purchased frequently by a large share of households and regarded as gateway goods for grocery retailers to attract consumers into stores.

Three major store formats are considered: A) value-oriented retailers (e.g. supercenters and price clubs) representing a super-cheap nontraditional shopping format characterized by low-pricing, broad assortment overall and especially in nonfood categories and low service; B) a format represented by traditional supermarkets and grocery stores, generally featuring promotional (HiLo) pricing, broad assortment in food categories and some service; C) high-end specialty stores (e.g. natural food supermarket chains) providing consumers with high-priced upscale product offerings and a higher level of service. To address the choices over all formats of retail outlets, we use a unique dataset collected by A.C. Nielsen, which covers the household purchases at any retail outlets including the retailers, such as Wal-Mart Inc. and Whole Foods Market Inc., that do not provide data to scanner data service firms. Our study is for a single large metropolitan area in a non-coastal U.S. city for a pair of two-year weekly samples, 2005-06 and 2007-08.

The remainder of the chapter is organized as follows. The next section contains an overview of the market background and trends, including the data, the consumer and retailer profiles and the consumption patterns of dairy products. Section 3 presents the model setup, estimation procedure and regression results for the consumption choice of organics. The last section contains a summary of our findings and their implications for marketing and farming decisions.

2. The market: Background and trends

2.1 The U.S. organic food market

Organic market has been one of the fastest growing markets in recent years. Aggregate organic food sales in the U.S. have maintained a 15-20% annual growth rate over the past decade. The report by (Organic Trade Association, 2009) indicates that the US sales of organic foods totaled nearly $23 billion in 2008, which marks a 15.8% increase compared to sales in 2007 and is over 6 times of the sales in 1997. The organic penetration rates, defined as organic food as a percent of total U.S. food sales, have increased from 0.97% in 1997 to 3.59% in 2009 (see figure 1). According to (The Hartman Group, 2008), over two-third of U.S. consumers buy organic products at least occasionally and about 28 percent of these organic consumers are weekly organic users.

Figure 1 also shows that the traditional supermarkets and value-oriented retailers have become more important outlets where consumers shop for organic food as their market shares combined have increased from 30% to 46% over the past decade. On the contrary, sales of organic foods through natural food chains, such as Whole Foods Market and Wild Oats, and other independent natural food stores peaked at 68% of total organic sales in 1995. By 2005, the market share of natural food channels had however dropped to 47% of sales.

2.2 The data

We use a multi-outlet panel dataset (Homescan by A.C. Nielsen) for a non-coastal U.S. city that covers a 208-week period between December 26, 2004 (hereafter January, 2005) and December 27, 2008. The dataset contains detailed purchase information for 6 food product departments (dry goods, frozen, dairy, deli, meat, and fresh produce) and over 600 product categories of food and non-food items sold in grocery stores or other retail outlets. The households report their purchases weekly by scanning either the Uniform Product Code (UPC) or a designated code for random weight products of all their purchases from grocery

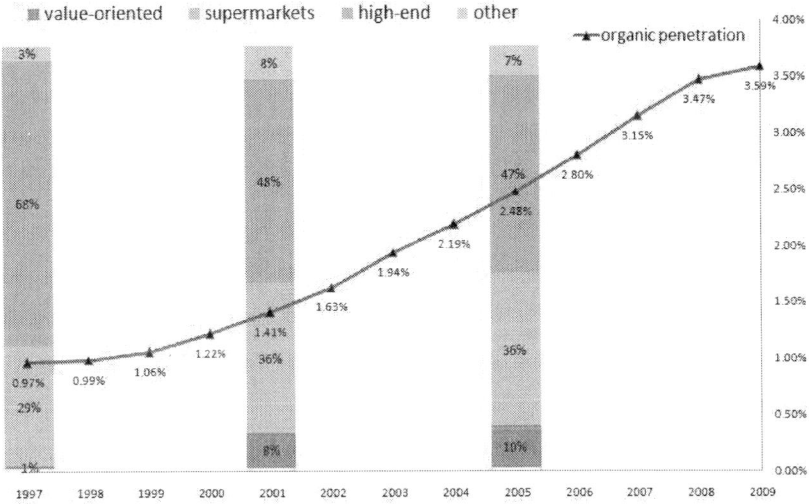

Fig. 1. The U.S. Organic Food Market, 1997-2009 (Organic Trade Association, 2007, 2009)

stores or other retail outlets. These purchase data include price, quantity, promotional information, and product characteristics. One of the product characteristics contained in the data is the identifier for organic products. For UPC-coded products, organic products can be identified by the presence of the USDA organic seal or with organic-claim codes created by A.C. Nielsen. For random-weight purchases, we use product descriptions to identify organic products.

	2005-06		2007-08	
	Mean	St Dev	Mean	St Dev
Number of households	710		942	
Number of shopping trips	161.34	101.80	137.83	86.75
Average spending per trip	23.06	24.67	18.40	20.44
Organic penetration rate (frequency)	1.20%	0.07	1.84%	0.10
Organic penetration rate (spending)	1.24%	0.08	1.93%	0.10
Household size	2.36	1.26	2.40	1.36
Income ($0000s)	6.33	3.64	6.86	4.12
Some college educated	87.9%	0.326	88.1%	0.324
Married	57.9%	0.494	58.0%	0.494
Preschool children (age <6)	5.8%	0.233	9.8%	0.297
School-age children (age 6-18)	21.2%	0.409	21.8%	0.413
Elderly (age >65)	22.5%	0.417	22.3%	0.417

Table 1. The Consumer Profile, 2005-2008

Due to the inconsistency on the coverage of random weight items over the analyzed period, we separate the four-year period into two, i.e. 2005-2006 and 2007-2008. The shopping-duration criterion was applied to ensure that each panelist was faithful in recording purchases and remained in the panel for the entire period. The resulting dataset had 710 households with a total of 45,877 shopping trips in 2005-06 sample and 942 households with 48,469 trips in 2007-08 sample. The selected retail chains for our analysis include 2 value-oriented retail chains consisting of 29 (37) stores, 4 traditional supermarket chains featuring 172 (147) stores, and 1 high-end specialty supermarket chain with 6 (7) stores in our 2005-06 (2007-08) sample.

2.3 The consumer profile

Descriptive statistics of the consumer profile are provided in table 1. The statistics show that there were significant reductions in shopping frequency and basket size over the two sample periods, which may indicate a greater reliance on food away from home during the latter period. Our data may also pick up some impact from the economic downturn for the U.S., particularly in the latter half of 2008 when the housing related credit crisis began to pick up steam. In this trend of consumption reduction, organic food is however relatively less affected as its share to total food consumption has increased from 1.20%/1.24% to 1.84%/1.93% in terms of frequency/spending (dollar amount). We observe no significant changes in household demographics, with an exception that the percentage of household with pre-school children (age<6) had increased from 5.8% (2005-06) to 9.8% (2007-08) on average.

2.4 The retailer profile

Table 2 depicts the characteristic differences among the retailers of three store formats. Location or network wise, high-end specialty stores are much less accessible compared to the other two formats as shown in number of stores, share of trips, share of spending, as

	2005-06			2007-08		
	value-oriented	super-markets	high-end	value-oriented	super-markets	high-end
Number of stores	29	172	6	37	147	7
Ave. travel distance (miles)	9.02	8.87	16.96	8.74	9.54	14.45
Share of trips	19.32%	79.46%	1.21%	21.47%	78.11%	0.43%
Organic% in total trips	0.27%	0.78%	25.07%	0.80%	1.38%	35.07%
Share of spending	18.49%	79.69%	1.81%	21.34%	78.01%	0.64%
Organic% in total spending	0.32%	0.96%	21.91%	1.02%	1.66%	29.99%
Pricing & Discount						
Price index (selected basket)	0.968	1	1.505	0.919	0.929	1.373
Organic PI (selected basket)	0.977	1	1.357	1.046	1.039	1.449
% discount (overall)	12.81%	40.12%	11.69%	10.25%	35.99%	9.51%
% discount (organics)	0.05%	0.29%	4.06%	0.08%	0.43%	3.42%
Broadness & Depth of Assortments						
Ave. broadness (# UPCs) per store	2038	1505	659	1557	1517	201
Organic% in total broadness	0.79%	2.28%	25.84%	1.35%	3.62%	31.84%
Ave. variety per category	33.98	63.72	9.07	32.86	57.78	4.68
Organic% in variety	7.47%	8.35%	49.54%	8.91%	10.52%	61.03%

Table 2. The Retailer Profile, 2005-2008

well as by the average travel distance from consumer's home to the store. However, it is documented that these high-end specialty stores are the major outlets for organic food, as their organic shopping rates are by far higher than those of the other two formats. In our selected sample market, traditional supermarkets remain the most important outlets among the three formats, although increasing market shares of value-oriented stores are observed in the data.

Regarding to pricing factors, we observe no significant price difference between value-oriented retailers and supermarket chains, but much higher prices at high-end specialty stores in both organic and non-organic alternatives. The data of discount use rates suggest that unlike the other two, traditional supermarkets promote promotional pricing. However, interestingly, we observed a much higher discount use rate applied to organic purchases at high-end stores than elsewhere. As to the coverage of product assortments, value-oriented retailers have broadest coverage but supermarket chains offer more varieties per category on average. The high-end specialty stores carried a much higher percentage of organic products in terms of both broadness and variety, but with a much small scale of assortments in general.

2.5 The consumption of dairy-case products

We select two staple dairy-case products, milk and eggs, as the center of our study. In our analyzed sample, milk was the most frequently purchased item in grocery shopping trips in both organic and conventional categories with shares of purchase frequency being about 20% and 3% respectively, while eggs ranked 9th (organic) and 10th (conventional) among all categories. In terms of dollar amount, the data (table 3) show that the expense shares were 5.5%~9.5% for milk and 1%~2.3% for eggs. As shown in table 3, we observe an increasing trend of organic penetration on both products - the share of organic food to total food expense increased from 6.0% to 10.3% for milk and from 0.6% to 1.3% for eggs. In addition, we observe significant drops in price premium of organic between the two periods of sample, which are likely to be associated with the market transitions that may have occurred due to Wal-Mart's market expansion in 2006.

3. The analysis: Consumption choice of organics

3.1 Data overview of consumption choice

Figure 2 depicts the consumption choice for milk and eggs based on actual purchase data recorded in our analyzed market during the period of 2005-2008. The data show a fast-growing consumption pattern of organics in the case of milk and eggs. 7.68% of milk

	milk		Eggs	
	2005-06	2007-08	2005-06	2007-08
Ave. share in total expense per trip	5.5%	9.5%	1.0%	2.3%
Ave. product expense per trip	0.93	1.24	0.18	0.30
Ave. % organic in total product expense	6.0%	10.3%	0.6%	1.3%
Ave. price premium (milk $/gallon, eggs $/dozen)	2.76	2.26	1.69	1.33
Ave. % discount used for purchase	22.8%	17.9%	27.6%	13.8%

Table 3. The Shopping Patterns of Milk and Eggs, 2005-2008

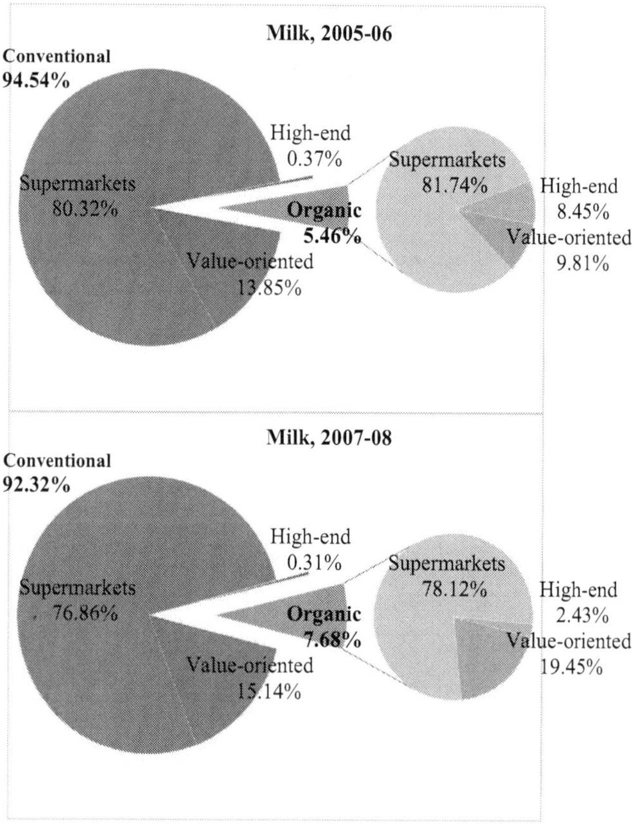

Fig. 2a. Consumption Choice by Store Format and Product Type for Milk, 2005-2008

purchase was organic in 2007-08, while organic milk purchase was only 5.46% out of total in 2005-06. In the case of eggs, organic choice though still accounts for only a small portion of egg purchase, its share has grown from 1.42% to 2.31%, which is over 60% of growth.

In terms of outlet choice, supermarket was the dominant store format of which consumers purchased their milk and eggs, accounting for 59.70% to over 80% of total number of transactions in all categories for both periods. We however observe a trend of market transition, in which consumers are switching their organic purchases from high-end specialty stores to value-oriented stores or supermarkets. In the case of milk, the value oriented retailers' share of organic milk doubled (increased from 9.81% to 19.45%) mainly at the expense of the high-end stores' sales: their share dropped from 8.45% to 2.43% between the two periods. This change reflects the marketing strategy by Wal-Mart and others to expand on organic offerings in 2006. The impacts are even more apparent in the market of organic eggs, as around 30% of consumers switched from high-end's to value-oriented stores and supermarkets for organic eggs purchase.

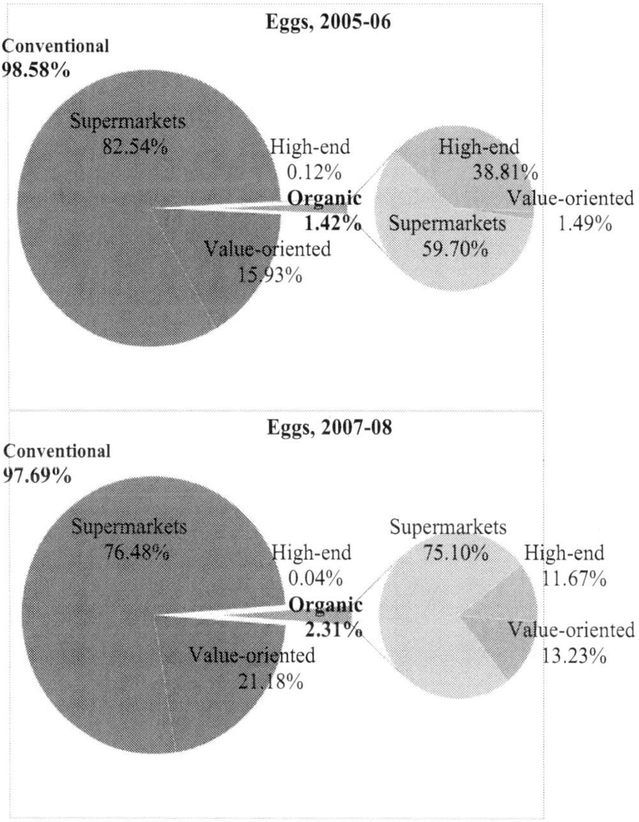

Fig. 2b. Consumption Choice by Store Format and Product Type for Eggs, 2005-2008

3.2 Econometric model specification

The choice between an organic versus a conventional food product is a typical binary discrete choice problem. Let U^o denote the utility of organic consumption and U^c that of conventional consumption. A common formulation of this kind of binary choice is the linear random utility model,

$$U^o = x'\beta_o + \varepsilon_o \text{and} U^c = x'\beta_c + \varepsilon_c. \tag{1}$$

$$\text{Prob}[Y=1 \mid x] = \text{Prob}\left[U^o > U^c\right] = \text{Prob}[x'\beta + \varepsilon > 0 \mid x], \tag{2}$$

where we denote by Y=1 the consumer's choice of organic product (o), $\boldsymbol{x}$ is a vector of the exogenous variables, $\boldsymbol{\beta} \equiv \boldsymbol{\beta}_o$-$\boldsymbol{\beta}_c$ is a vector of parameters (organic against conventional), and ε is a random error. In this chapter, we adopt the logit model setup, i.e. assuming that the probability follows the logistic distribution,

$$\text{Prob}[Y=1|x] = e^{\mathbf{x}'\boldsymbol{\beta}} / \left(1+e^{\mathbf{x}'\boldsymbol{\beta}}\right) + \varepsilon = \Lambda(x'\beta) + \varepsilon, \tag{3}$$

where Λ denotes the logistic cumulative distribution function.

The elements of exogenous variable set (x) include the price premium of organic versus conventional products in the store at trip t (price premium), the discount use rate (% discount), days between trips, distance between consumer's home and store (distance), number of stores within the shopping range by format, total spending and organic percentage in other items purchased in the same trip, income and household demographics. The set of demographic characteristics includes household size and dummy variables identifying the percentages of 1) college educated householder, 2) married householder, 3) family with preschool children, 4) family with school-aged children, and 5) the elderly. In addition, we include household's format loyalty to three formats alone and their interaction terms with price premium variable to allow for differential price sensitivity and fixed component of utility to differ among households with differentiated preferences over formats. In addition, we employ the same set of demographics described above for preference heterogeneity.

In sum, the complete empirical specification of organic choice model is as follows:

$$\begin{aligned}\text{Prob}[Y=1|x] = \Lambda(&\Sigma_f \beta_{1f}\textbf{loyalty}_f + \Sigma_f \beta_{2f}(\textbf{price premium} * \textbf{loyalty}_f) \\ &+ \beta_3\%\textbf{discount} + \beta_4\textbf{days between trips} \\ &+ \beta_5\textbf{distance} + \beta_6\textbf{total spending of the trip} \\ &+ \beta_7\%\textbf{organic in other items of the trip} \\ &+ \beta_8\textbf{income} + \beta_9\textbf{household size} \\ &+ \beta_{10}\textbf{college educated} + \beta_{11}\textbf{married} \\ &+ \beta_{12}\textbf{preschool children} + \beta_{13}\textbf{school}-\textbf{age children} \\ &+ \beta_{14}\textbf{elderly}) + \varepsilon.\end{aligned} \tag{4}$$

Definitions and measures of variables are summarized in table 4 and further discussed in next subsection.

3.3 Estimation procedures and measures

Within each set of two years, we use the first 26 weeks as our "initialization" period to identify shopper types and format-specific indexes to avoid potential endogeneity between quality and store format choices. The remaining 78 weeks were used as the "estimation" sample. The estimation is based on every shopping trip of households with shopping duration being no longer than 30 days during the estimation period at seven major retail chains in the market.

Format Loyalty. The format-specific loyalty for a household is represented by the percentage of trips that the household made to the format during the initialization period. Specifically, we use the following standard measure of loyalty (*FL*) used also by (Bell, Ho, and Tang 1998; Briesch, Chintagunta, and Matzkin 2010):

$$FL_{h,f} = \left(NV_{h,f} + 0.5\right) / \left(\Sigma_f NV_{h,f} + 1\right) \text{ for} f = A,C \tag{5a}$$

and

$$FL_{h,B} = 1 - FL_{h,A} - FL_{h,C}. \quad (5b)$$

where $NV_{h,f}$ is the number of visits to format f stores by household h during the initialization period. This index reveals the shopper's preference toward a specific format due to probably the familiarity about the store layout, the general prices and assortments, and the convenience and quality of service, based on his/her past shopping experience.

% Discount. We use household discount use rates, calculated from the household purchase information during the initialization period, to capture their preference between promotional pricing (HiLo) and everyday low pricing (EDLP). We expect a household with a high rate would prefer the format in which stores/chains use HiLo pricing instead of EDLP, and otherwise for low-discount-use households.

Variable	Definition	2005-6	2007-8
	Milk		
Y (choice)	1 if organic, 0 if conventional	1.94	1.92
price premium	organic price – conventional price ($)	0.01	0.01
%discount	1 if any discount (sale or coupon) applied, 0 otherwise	0.23	0.18
%organic in other	% of organic/total expense in other items of the trip	0.23	0.19
	Eggs		
Y (choice)	1 if organic, 0 if conventional	1.98	1.98
price premium	organic price – conventional price ($)	0.23	0.19
%discount	1 if any discount (sale or coupon) applied, 0 otherwise	0.27	0.15
%organic in other	% of organic/total expense in other items of the trip	0.29	0.29
loyalty (value-oriented)	% of trips that household made to the format (value-oriented, supermarkets, high-end) during the initialization period	0.18	0.22
loyalty (supermarkets)		0.81	0.77
loyalty (high-end)		0.01	0.01
days between trips	number of days between two shopping trips	5.05	5.63
distance	the distance between consumer's home and store	9.16	9.39
total spending	total transaction amount recorded for the shopping trip	23.10	16.99
income	household income (in $1,000)	6.38	6.89
household size	number of persons in the household	2.34	2.39
college educated	1 if householder is college educated, 0 otherwise	0.89	0.89
married	1 if married householder, 0 otherwise	0.57	0.57
preschool children	1 if family has child(-ren) aged <6, 0 otherwise	0.06	0.11
school-age children	1 if family has child(-ren) aged 6 ~18, 0 otherwise	0.23	0.24
elderly	1 if householder is aged 65 and above, 0 otherwise	0.24	0.23

Table 4. Description of Variables for Consumer Panel Households, 2005-2008

Rank	Description of Product Category
1	DAIRY-MILK-REFRIGERATED
2	BAKERY - BREAD - FRESH
3	CEREAL - READY TO EAT
4	SOFT DRINKS - CARBONATED
5	YOGURT-REFRIGERATED
6	FRUIT
7	SOUP-CANNED
8	COOKIES
9	VEGETABLES
10	EGGS-FRESH
11	PRECUT FRESH SALAD MIX
12	CANDY-CHOCOLATE
13	FRUIT DRINKS-OTHER CONTAINER
14	WATER-BOTTLED
15	BEEF
16	SNACKS - TORTILLA CHIPS
17	FRESH CARROTS
18	FRESH STRAWBERRIES
19	FRESH FRUIT-REMAINING
20	RICE - MIXES
21	FRUIT JUICE - APPLE
22	FRUIT JUICE-REMAINING
23	PREPARED FOODS
24	YOGURT-REFRIGERATED-SHAKES & DRINKS
25	FROZEN FRUITS
26	VEGETABLE JUICE AND DRINK REMAINING
27	MEAT PRODUCTS-IMITATION & ADDITIVES
28	FISH
29	WHIPPING CREAM
30	SEAFOOD-SHELLFISH

Table 5. Base Basket for Price Comparison among Store Formats

	Store Format	Value-oriented		Supermarkets		High-end	
	Shopper Type	Organic	Conv.	Organic	Conv.	Organic	Conv.
Milk	2005-06	18.81%	31.67%	38.07%	50.59%	56.46%	70.69%
	2007-08	19.92%	24.80%	30.03%	39.52%	51.51%	59.49%
Eggs	2005-06	275%	282%	265%	286%	290%	175%
	2007-08	106%	118%	132%	138%	233%	139%

Table 6. Organic Price Premium by Store Format and Shopper Type of Actual Purchase for Milk and Eggs, 2005-2008

Price Index. To generate the format-specific price index, we first select a comparable basket of items (30 product categories, see table 5 for the details) available for all three formats. After the basket is constructed, we then calculate the average household consumption pattern for the selected product categories in the basket from the initialization sample. Using these base quantities together with the format-specific category price indexes, we estimate the cost at each format, which we refer as overall price index.

Price Premium. Since we observe only the prices for the products chosen by the household, we use the following procedure to recover the "missing prices," i.e. the ones for the alternative choice, and then construct the price premium of organic based on the price difference between the two. First, we look for the prices for the alternative at 1) the same store, 2) the top 3 stores that the household most frequently visited in the past, 3) the same store chain, and 4) the format of the specific store. We then use the average prices from the most relevant group of stores (i.e. in the above order) as the proxy for the missing prices of the alternative choice.

Table 6 summarizes the average price premium of organic versus conventional product (milk and eggs) by store format and shopper type based on the actual purchase of each transaction. The data show that price premium varies among stores of different formats and between organic and conventional shoppers. In particular, we observe that organic price premiums are at minimum in value-oriented stores, while high-end stores feature much higher organic price premiums. In addition, consumers who purchased organic products in general face the lower organic price premium compared to those who purchased conventional alternative at the outlets of the same store format, except for the case of eggs at high-end stores. We also observe sizably diminishing organic price premiums for all outlets over the two periods. For example, in the case of eggs, the organic price premiums in the value-oriented stores dropped from 275% to 106%, which is less than half of the former. The only exception is the case of organic milk purchased in the value-oriented stores, the price premium was 18.81% in 2005-06 and 19.92% in 2007-06. It likely indicates that organic price premium for milk may have reached the low-end retailers' pricing constraint bounded by a certain level of markup above the high production costs of organics.

3.4 The regression results

Table 7 presents the parameter estimation results from maximum likelihood estimation (MLE) regression. Several key findings emerge from our analysis. The first three rows of table 7 reveal the statistical association between store loyalty and the likelihood of purchasing organics. The results are quite mixed yielding no clear conclusion about a discernable pattern of behavior. For example, consumers with higher loyalty to value-oriented stores were less likely to purchase organic milk in both periods. However, for eggs increasing loyalty did not affect the probability of purchasing organic in the early period (insignificant parameter estimate) but increased the probability of purchasing organics in the latter period. For supermarkets, increased loyalty is associated with a lower likelihood of purchasing organics in all periods for both milk and eggs. For high-end stores, increased loyalty led to an increased likelihood of purchasing organic milk in the early period but decreased probability in the latter period. Increased loyalty to high-end stores had no impact on organic egg purchasing.

When we look at loyalty with respect changes across the two time periods, there is a general pattern of increased organic purchasing in formats that have increased their organic offerings. Note that in the case of value-oriented formats and supermarkets, the parameters

	Milk		Eggs	
	2005-06	2007-08	2005-06	2007-08
loyalty (value-oriented)	-2.8573**	-1.2704**	-4.5387	3.9902**
	(0.3431)	(0.2358)	(5.6458)	(1.4506)
loyalty (supermarkets)	-1.8855**	-2.2079**	-3.7655**	-2.8702**
	(0.1720)	(0.1708)	(0.9297)	(1.0625)
loyalty (high-end)	9.3185**	-4.8025**	-7.3624	-4.8729
	(0.9229)	(1.1015)	(5.2753)	(3.5255)
price premium* loyalty	-40.5001**	-4.1765	-5.9651	-32.2421**
(value-oriented)	(8.9784)	(6.0441)	(24.8313)	(9.4999)
price premium* loyalty	-7.8783**	-3.3919	4.8699	11.6465**
(supermarkets)	(2.9340)	(3.7701)	(3.8133)	(3.3894)
price premium* loyalty	105.6417*	76.3539*	43.8518	16.6296
(high-end)	(44.9088)	(35.2421)	(23.4916)	(11.7079)
% discount	-0.1066	-0.2957*	-0.4048	-1.7209*
	(0.1436)	(0.1385)	(0.5430)	(0.7652)
days between trips	-0.006	0.0179*	-0.0651	0.0254
	(0.0118)	(0.0074)	(0.0456)	(0.0203)
distance	0.0018	0.0060**	-0.0884**	-0.0305**
	(0.0024)	(0.0021)	(0.0303)	(0.0118)
total spending of the trip	-0.0075**	-0.0021	-0.0045	-0.0260**
	(0.0022)	(0.0018)	(0.0041)	(0.0073)
% organic in others items	0.1871**	0.2725**	0.3118**	0.4334**
of the trip	(0.0302)	(0.0323)	(0.0650)	(0.0512)
income	0.0179	0.0025	0.2926**	0.0498
	(0.0148)	(0.0106)	(0.0781)	(0.0276)
household size	0.1850**	-0.0918	-1.4851**	-1.0755**
	(0.0618)	(0.0498)	(0.4480)	(0.2403)
college educated	-0.9139**	-0.0951	1.0390*	-0.8338*
	(0.1410)	(0.1386)	(0.5193)	(0.3437)
married	0.0433	0.2235	-0.8613	0.2264
	(0.1174)	(0.1254)	(0.5706)	(0.3853)
preschool children	0.2939	0.3032	2.7129**	0.6898
	(0.2139)	(0.1585)	(0.7454)	(0.3979)
school-age children	-0.4939*	-1.0582**	-0.2508	0.0025
	(0.2209)	(0.1936)	(0.4016)	(0.5546)
elderly	-1.6509**	-1.7207**	-2.4469**	-2.1501**
	(0.2563)	(0.1736)	(0.6518)	(0.4944)
number of observation	13206	15685	4732	6750
log likelihood	-2895.0943	-4149.0844	-336.5629	-563.9128
Wald statistics	2080.53	2680.72	776.76	989.14
(Prob>χ^2)	(<0.001)	(<0.001)	(<0.001)	(<0.001)
Pseudo R^2	0.6837	0.6184	0.8974	0.8795

Note: The "conventional" is the base outcome for all variables. Robust standard errors in parentheses; *, ** denote statistical significance at 5% and 1%, respectively.

Table 7. MLE Parameter Estimates for Choice of Organic Milk and Eggs on loyalty are a) higher in the latter period compared to the former and b) statistically significant. This result seems to suggest a pattern of increasing acceptance among loyal shoppers that previously may not have sought out organics in the high-end stores. Additionally, increased loyalty in milk purchases in high-end store was highly associated with increased organic purchase in the early period while highly associated with decreased organic milk purchases in the latter period. This result suggests that shoppers loyal to high-end stores are either increasing the purchase of non-organic milk in the latter period or have shifted organic milk purchases to different formats.

As a second finding, we observe a differential pattern associated with the impact of price premiums that interact with loyalty. As the price premium for organic eggs rise, it significantly mitigates the impact of loyalty on purchases in the value-oriented format. However, we observe the exact opposite pattern on the impact of loyalty within the supermarket format: higher price premium correlate to an enhanced impact of loyalty on the likelihood of purchasing organic. The results suggest strongly the presence of quite different consumer attitudes about organic eggs across formats. In the value-oriented stores, loyalty and price consciousness seem to go "hand-in-hand" whereas in supermarkets, loyal consumers appear to have "switched" to perceive premiums as a measure of egg quality and are willing to pay the extra funds to obtain the brands with highest actual or perceived qualities. A similar pattern is noted for milk but the switch now occurs between the supermarket and the high-end store format: price premiums in the value-oriented and supermarket (high-end) formats mitigate (enhance) the impact of loyalty on the likelihood of purchases. Note, however, that this switching pattern tends to disappear in the latter period for milk while it is only present in eggs during the latter period. These different patterns are reflective of the rapid changes occurring in organic offerings and in the various rates of acceptance about organics across the entire food-at-home market.

Figure 3 and table 8 provide some examples to demonstrate the differences in fixed tendency (constant term) and in price sensitivity among households with different format loyalty tendency. Among the extreme cases of (A, B, C), i.e. households with 100% loyalty to single store format, A(value-oriented) is most price sensitive, while C(high-end) is least price sensitive. Type D(value-oriented+supermarkets) households feature negative price sensitivity for both milk and eggs in both periods, while other combinations do not necessarily yield negative price sensitivity. Using the mean values of format loyalty from our 2005-06 (H1) and 2007-08 (H2) samples, we find that the average household's fixed tendency (constant term) to purchase organic over conventional milk was -1.90 in 2005-06 and -2.03 in 2007-08, implying conventional milk was the dominant choice with about 7.389 odds ratio versus organic alternative.[1] This average household has negative price sensitivity -12.12 in 2005-06 and -2.67 in 2007-08, meaning that a unit of price premium increase would reduce the odds ratio of organic over conventional by a larger amount in 2005-06 than that in 2007-08. For this average household, we however find a positive price coefficient in the case of eggs consumption, suggesting the higher price premium, the more likely the household would purchase organic eggs. This may be a result of differentiated perceived product quality. As already discussed, consumers may see the higher price of organic eggs as an indicator of better quality (i.e. Cadillac pricing).

---

[1] The odds ratio of organic over conventional milk is about exp(-2)=0.1353, of which the inverse ratio is 7.389.

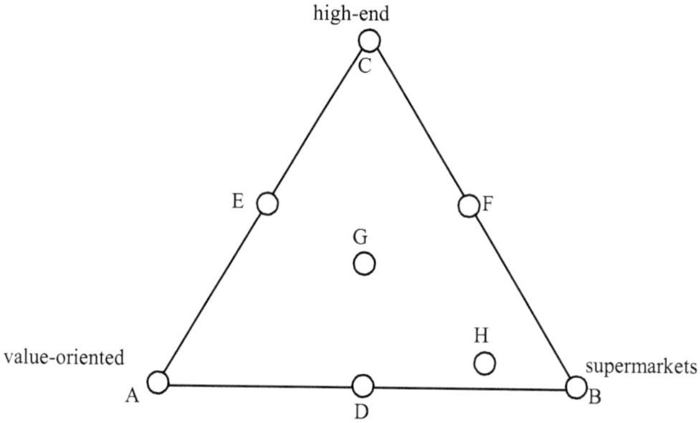

Fig. 3. Example Households with Specific Format Loyalty Tendency Combination

	Loyalty			milk				Eggs			
	value-oriented	super-markets	high-end	2005-06		2007-08		2005-06		2007-08	
				β1	β2	β1	β2	β1	β2	β1	β2
A	1	0	0	-2.86	-40.50	-1.27	-4.18	-4.54	-5.97	3.99	-32.24
B	0	1	0	-1.89	-7.88	-2.21	-3.39	-3.77	4.87	-2.87	11.65
C	0	0	1	9.32	105.64	-4.80	76.35	-7.36	43.85	-4.87	16.63
D	0.5	0.5	0	-2.37	-24.19	-1.74	-3.78	-4.15	-0.55	0.56	-10.30
E	0.5	0	0.5	3.23	32.57	-3.04	36.09	-5.95	18.94	-0.44	-7.81
F	0	0.5	0.5	3.72	48.88	-3.51	36.48	-5.56	24.36	-3.87	14.14
G	0.33	0.33	0.33	1.53	19.09	-2.76	22.93	-5.22	14.25	-1.25	-1.32
H1	0.18	0.81	0.01	-1.90	-12.12			-3.96	3.48		
H2	0.22	0.77	0.01			-2.03	-2.67			-1.37	1.96

Table 8. Differences in Fixed Tendency and Price Sensitivity among Example Households

Returning to table 7, we find that the households who prefer "discounts" or with larger basket sizes tended to purchase more conventional milk and eggs. Unsurprisingly, we find higher likelihoods for households that purchase other organic foods and with higher incomes to purchase organic milk and eggs. We did not find a clear indication for the impact of days between trips on organic choice. Households that traveled further to shop for food tended to favor the purchase of conventional eggs.

The coefficients of the remaining demographic variables show mixed results in general. Those consumers with a college degree tended to purchase fewer organic milk and eggs except in the first period for eggs. The consumers' marital status shows no statistical impact as being a differentiating factor in their consumption choice of organics. We however find that organic eggs are more attractive to small households. In addition, families with preschool children are more likely to buy organics, while families with school-age children and the elderly households are more likely to choose conventional over organic milk or eggs. This is an interesting finding as it seems to suggest that the younger generation (parents) are more likely to consume organics, which is consistent with findings in many studies and marketing efforts focusing on organic infant foods and foods to attract the younger generation.

4. Conclusion

In this chapter, we study the determinants of consumption choice between organic and conventional alternatives for two staple foods, milk and eggs. In addition to prices, consumer shopping behavior and demographic characteristics, we incorporate store format choice into our analysis to reflect variation in consumption choice among households patronizing different store formats. Using A.C. Nielsen Homescan data, we examine three store formats, value-oriented, supermarkets and high-end, for the duration of periods of recent major market transitions in the U.S. organic food markets.

The Homescan data recorded from our analyzed market confirm three common observations on consumption choice of organics for milk and eggs: 1) a rapidly growing demand for organics, 2) a trend of organic market transition from high-end specialty stores to general store formats, especially value-oriented stores, and 3) a sizable reduction in organic price premium for all retail outlets. Several key findings emerge from our regression analysis. First, we find much statistical support that the degree of loyalty within each format matters in terms of making organic choices. However, the patterns of association are very different and depend critically on the product, the time-period, the format being studied, and the sensitivity of price premiums on organic products. Our results strongly suggest that the market for organic foods remains very dynamic and inherently risky for upstream suppliers trying to gauge proper levels of commitments to organic supply. Unsurprisingly, we observe considerable price sensitivity to organics in the value-oriented formats suggesting that low price premiums are required to stimulate demand in these stores. Additionally, we found evidence of "Cadillac pricing" in the latter period for eggs: higher organic price premiums are associated with an increased likelihood of an organic purchase. We also find income, families with preschool children and organic penetration rate in other items purchased at the same shopping trip have positive impacts on choosing organic milk and eggs over their conventional alternatives. Finally, we find mixed results from other demographic factors, except that the elderly prefer conventional to organic on milk and eggs consumption.

In sum, we have presented a close connection between store format choice and consumption choice of organics. We show that store format choice, prices, as well as quality perception are important to consumer's choice between organic and conventional food products. We have also documented the evidence from the actual purchase data on the impacts of recent economy down turn as well as marketing expansion made by Wal-Mart and others. These results provide useful insights for farmers and retailers in their marketing and developing decisions on organic agriculture.

5. Acknowledgment

We thank Ephraim Leibtag and Biing-Hwang Lin at Economic Research Service of the U.S. Department of Agriculture, and the A.C. Nielsen Company for supplying key data for this study.

6. References

Bell, D. R., T. Ho, & C. S. Tang. (1998). Determining Where To Shop: Fixed and Variable Costs of Shopping. *Journal of Marketing Research* 35 (3), pp. 352-369.

Bell, D. R., & J. M. Lattin. (1998). Shopping Behavior and Consumer Preference for Store Price Fromat: Why "Large Basket" Shopping Prefer EDLP. *Marketing Science* 17 (1), pp. 66-88.

Bhatnagar, A., & B. T. Ratchford. (2004). A Model of Retail Format Competition for Non-durable Goods. *International Journal of Research in Marketing* 21, pp. 39-59.

Briesch, R. A., P. K. Chintagunta, & R. L. Matzkin. (2010). Nonparametric Discrete Choice Models With Unobserved Heterogeneity. *Journal of Business and Economic Statistics* 28 (2), pp. 291-307.

Ho, T., C. S. Tang, & D. R. Bell. (1998). Rational Shopping Behavior and the Option Value of Variable Pricing. *Management Science* 44 (12), pp. 145-160.

Hsieh, M.-F. (2009). A Theory of Consumer Format Choice and Organic Penetration. Paper presented at Southern Economic Association 79th Annual Meetings, November 22, 2009, at San Antonio, TX.

Hsieh, M.-F., & K. W. Stiegert. (2012). Store Format Choice in Organic Food Consumption. *American Journal of Agricultural Economics* (Proceedings Issue).

Kavilanz, P. B. (2008). The High Price of Going 'Organic'. *CNNMoney.com.*

Messinger, P. R., & C. Narasimhan. (1997). A Model of Retail Formats Based on Consumers' Economizing on Shopping Time. *Marketing Science* 16 (1), pp. 1-23.

Organic Trade Association. (2007). The OTA 2007 Manufacturer Survey Overview. Greenfield, MA: Organic Trade Association.

Organic Trade Association. (2009). The OTA 2009 Organic Industry Survey. Greenfield, MA: Organic Trade Association.

Tang, C. S., D. R. Bell, & T.-H. Ho. (2001). Store Choice and Shopping Behavior: How Price Format Works. *California Management Review* 43 (2), pp. 56-74.

The Hartman Group. (2008). The Many Faces of Organic 2008. Bellevue, WA: The Hartman Group, Inc.

2

Determinants of Purchasing Behaviour for Organic and Integrated Fruits and Vegetables: The Case of the Post Socialist Economy

Aleš Kuhar[1], Anamarija Slabe[2] and Luka Juvančič[1]
[1]*University of Ljubljana, Biotechnical Faculty*
[2]*Institute for Sustainable Development*
Slovenia

1. Introduction

Modern food consumer is highly concerned about the safety and quality of the food products purchased. This concern goes simultaneously with their awareness of the relation between the production practice and quality of food products, as well as environmental concern in regards to food (Thøgersen & Ölander, 2002). Moreower, the awareness has contributed towards growing demand for food from non-conventional production practices as well as an increasing consumer interest in having a closer relationship with the food producer (Thompson, 1998; Wier et al., 2003; Vermeir & Verbeke, 2006; Botonaki et al., 2006). This change has been especially significant in the demand for organic foods, since the global annual organic sales are estimated at around 38.6 billion US Dollars in 2006 which is double figure in comparisons with the figure in 2000 (Willer et al., 2008). Consumption of organic food is highly concentrated in North America and Europe since these two regions comprise 97% of the global demand (Sahota, 2008). Consumption in these two markets is growing at close to 20% annually (Wier and Calverley, 2002; Halberg et al., 2006). Fruit and vegetables is the largest segment in the European organic food market with almost one forth of total organic food sales in 2006. Moreover, organic fruit and vegetables represented about 2 percent of all fruit and vegetable sales in Western Europe (Willer et al., 2008). This pattern is related particularly with the increased awareness of the importance of a healthy diet and positive perception of fruit and vegetable in this respect (Connor, 1994; Viaene et al., 2000; William & Hammit, 2001, Lambert N. 2001; Belows, et al., 2008).
Response towards these trends on the supply side is also evident, since organic agriculture is one of the most rapidly developing market segments in both developed and developing countries (Halberg, 2006). Land area under organic agriculture has increased from 16.9 million hectares in 2000 to 30.4 in 2005 globally, whereas the growth in Europe was even faster (Willer et. al. 2008). In 2005 the organic area made up 3.9% of the total utilised agricultural area in the European Union, and the highest proportions were recorded in Austria with 11.0%, Italy with 8.4%, and the Czech Republic and Greece both with 7.2% (Eurostat, 2007). Simultaneously the global organic food chain has been transformed from a local network of producers and consumers to a highly coordinated and formally regulated supply system (Raynolds, 2004).

Agricultural policies in many developed countries, including European Union, have responded to favourable market trends and benefits external to the markets (e.g. environmental and spatial impacts) arising from organic farming and other sustainable agricultural practices (Hamm et al., 2002). This is reflected in rising importance of measures to encourage and promote organic farming (OECD, 2003). With new EU legislation geared towards increasing the production of organic food in Europe applied since 2009, the growing awareness of organic food and its benefits should see the market continuing its high growth into the near future.

There is a rather great amount of research work attributed to the attitudes of consumers towards safe food, both in broad sense and with a particular accent towards organic fruit and vegetables. Determinants of food choice and radical changes of related behavioural patterns are challenging and important in many aspects. Initially, the main focus was to investigate consumer's needs and motivations in order to support agro-food industry and retailing sector in searching competitive advantage with supreme supply. Currently the questions of food choice became also an issue from the perspective of public health and motivation of the policy makers in developed economies to improve dietary patterns of the population. Contemporary research literature on food choice considers product attributes as one of the perspectives to increase understanding of consumer or buyer (Assael 1998). A product is comprehended as an aggregation of several characteristics and components - referred as product attributes; upon which buyers makes their choices. Consumers during a complex cognitive process form beliefs and develop attitudes and intentions. A number of papers have dealt with the consumer behaviour, decision-making process and attitudes towards notion of safety related to food, both in broad sense and with a particular accent towards food produced under a specific quality assurance system like organic agriculture.

Previous studies showed that consumers perceive organic food as of higher quality, safer and fresher (e.g. Thompson & Kidwell, 1998; Schifferstein & Oude Ophuis, 1998; Loureiro et al., 2001; Botanaki et al., 2006; Kihlberg, I. & Risvik, E. 2007). Another dimension of attributes related to organic food is positive environmental impact, since it is perceived as produce grown as natural and without chemicals (Grunert & Juhl, 1995). However consumers' concerns regarding the pollution tend to be less important drivers for organic food consumption than so called private benefits (Weir et al., 2003; Bellows et al., 2008). Therefore healthiness of the products in comparison to conventional food options is among the main reasons for organic food purchase (Loureiro et. al, 2001; Krystallis & Chryssohoidis, 2005; Kihlberg & Risvik, 2007; de Magistris & Gracia, 2008). In this respect the concern for children healthy diet has also been identified (Latacz-Lohmann and Foster, 1997). Yiridoe et al. (2005) exposed the importance of knowledge on organic food products as a factor that is strongly affecting buying decision, since consumers without information cannot differentiate the attributes of organic from conventional alternatives. Related factor to knowledge is trust in system of labelling and conformity to standards of production practices defined in regulation (Botonaki, 2006; Achilleas & Anastasios, 2008). The importance of price as a barrier to purchase fruit and vegetable from non-conventional production systems is confirmed by an increasing amount of research that assess the consumers' willingness to pay a premium for organic or safe products (e.g. Weaver et al. 1992; Underhill & Figueroa, 1996; Govindasamy & Italia, 1999; Boccaletti & Nardella, 2000; Canavari et al., 2005; Batte et al., 2007). Production yields are considerably lower for organic production and therefore achieved price premium is a key determinant for organic farming

attractiveness and profitability. Consumers' willingness-to-pay a premium shows the value they place on the product attributes, whereas socio-demographic characteristics, perceived quality and risks determine the value consumers are willing to pay.
From the brief literature review it is evident, that consumer behaviour in relation to quality identified food such as organic and food from integrated production system is an evolving phenomenon, and therefore needs to be constantly studied. Only precise knowledge regarding the consumer perception in this respect will provide sound foundation for business development strategies of agro-food producer. Likewise, this information is needed also to assist rapidly emerging food and agricultural policies that prevalently place stimulation of high quality fruit and vegetables consumption as an important objective. Therefore, a rapid growth in demand and production of these food categories necessitate continuous research in order to document and understand the evolution of the markets. Necessity to investigate consumers' attitudes towards quality identified food categories is even more expressed in case of countries where the corresponding markets have emerged only relatively recently. In these cases the underlying knowledge regarding consumer attitudes, perception and behaviour in relation to organic food is also rather insufficient. Countries that acceded to the European Union are a good example of such markets, since the accession brings inclusion into the common policy framework where organic agriculture and other quality identified food production play important role. In this chapter we are trying to contribute to a better acquaintance with consumers' attitudes and perception towards organic and integrated fruit and vegetables in the case of Slovenia. First, we describe the process of development of the organic and integrated food production systems, which is followed by the presentation of the Slovenian organic food market volume and the corresponding sales channel structure. Next, the detailed analysis of price premiums for organic food products at the retail level in major marketing channels is presented, which confirms that the Slovenian organic food market is still immature. The last part of the chapter is designed to contribute to a better acquaintance with consumers' attitudes and perception towards organic and integrated fruit and vegetables in the case of Slovenia. A country-wide survey has been conducted in order to develop a consumer behaviour model of qualitative choice which elucidates and quantifies the impact of various determinants influencing purchasing behaviour of organic and integrated fruit and vegetables in Slovenia. Results of this research are aimed at enabling more effective marketing strategies of organic and integrated fruit and vegetable producers in Slovenia, but also to support public policy initiatives to stimulate demand of these categories of food

2. Development of the organic and integrated food production system in Slovenia

2.1 A general overview

Market for organic fruit and vegetables in Slovenia started to develop in the late nineties, whereas the first attempts to promote integrated production were present a decade earlier (MAFF, 2006). Foundations for development of adequate certification system started with the establishment of Slovenian Organic Farmers' Association in 1997 by market oriented organic farmers. They were motivated from the cooperation with merchants and they both wanted to be able to put certified Slovenian products on the market, in order to satisfy evolving consumers demand, and to protect themselves from fraud and false organic labelling. The most important task of the association was firstly, to develop organic control and certification, and secondly, to promote certified organic products and to support development of organic farming. Standards were than prepared by an NGO Institute for Sustainable Development and were published also by the Slovenian Ministry of Agriculture Forestry and Food. The standards were accordant to the IFOAM Basic Standards and some other national standards (e.g. Austrian Ernte and German Bioland). As early as in 1999, a total of 300 farms applied for certification (Slabe, 2002). Since then controlled farming systems such as integrated and organic farming are on the increase. During the period prior the Slovenian accession to the European Union national regulations have been accepted regarding to organic farming and integrated agricultural production system. After the accession Slovenia adopts entire EU system of food quality identification including organic farming (e.g. EEC. 2092/91). In the year 2005 "Action Plan for Development of Organic Farming in Slovenia until 2015" was adopted by the government (MAFF, 2005). In spite of the lack of market-related data at that time, one of its goals was "By 2015 a 10% share of organic foodstuffs of Slovenian origin on the national market is to be achieved." However, at that time, the lack of data on organic food market and especially on its share of the total food market was a general feature even in the EU member states with more developed organic sectors (Padel S. et al., 2008). Today a significant part of fruit production (especially apples and pears) is produced following the integrated standards and marketed under the national label and two private collective marks, one for fruit and the other for vegetable.

In 2010 around 58 thousand hectares of farmland was cultivated under the integrated farming system which represents more than one fourth of total arable land and permanent plantations in Slovenia (MAFF, 2011). In total 5.576 farms acquired an integrated farming certificate for the same year. More than two thirds of the area is arable land; mainly maize and feed grain and therefore the crop enter food chain as animal feed. The rest of the area is under permanent plantations including vineyards and particularly the produced fruit and vegetable is differentiated on the market.

	2000	2001	2002	2003	2004	2005	2006	2007	2008	2009	2010
Area in organic control (ha)	5.446	10.828	13.828	20.018	23.019	23.169	26.831	29.322	29.836	29.388	30.735
Share of organic land in total UAA	1,1%	2,1%	2,7%	3,9%	4,7%	4,6%	5,5%	5,9%	6,1%	6,3%	6,4%
No. of holdings in organic control	600	1.000	1.160	1.415	1.582	1718	1.876	2.000	2.067	2.096	2.218
No. of certified holdings	115	322	412	632	910	1.220	1.393	1.610	1.789	1.853	1.897

Source: Ministry of Agriculture, Forestry and Food 2011; SORS, 2010.

Table 1. Area of agricultural land in organic control and its share in UAA, No. of holdings in organic control and no. of certified holdings in Slovenia, 2000 – 2010;

On the other hand organic production is still rather sporadic and the market presence is rather low. In 2010 the area under organic farming was almost 31 thousand hectares, but almost ninety percent of the land is grassland and pastures. Only minor part of that area is intended for differentiated market production, since organic animal products (e.g. meat or dairy) are extremely rare. Rather, the produce is entering conventional supply chains, and the prevalent motivation of the farmers to enter the organic control is to be eligible for additional budgetary support. This means that the development of organic farming was not

predominantly market driven and therefore not entirely related to the consumer demand. The share of organic land in total utilised agricultural area was rapidly growing in the last decade and has reached 6,4% in the year 2010. In the same year 2,218 farms have been in the system of organic control and 1,897 have acquired the organic farming certificate (MAFF, 2011). It should be noted that intensity of growth for all the indicators of organic farming has slowed down in the last few years, which is mainly due to the already mentioned "passive organic farmers phenomenon" and therefore low market orientation.
There is still only limited marketing information available for the Slovenian organic sector, despite a stable increase of consumers demand and development of the supply. Some fragments of market-related estimates can be found in the EU research project OMIaRD (Hamm U. et al., 2002; Hamm U. and Gronefeld F., 2004). However, the estimates provided within the project largely lacked local expert verification and can thus have only an indicative value. The first exploratory analysis of the organic food supply was done by Slabe et al. (2005) who revealed significant organisational weaknesses of the domestic production and processing supply chain. The main identified drawback is insufficient supply, especially for the most demanded products such as fresh vegetables and fruits, grain and processed vegetables. Furthermore, there were no producer organisations and hence deficient marketing capabilities as well as fragmented production capacities. The problem of non-differentiation of the organic cattle was identified, since the considerable part of the farmers was selling their organic animals as conventional. However, the range of domestic organic food products on the Slovenian market is relatively broad, but the quantities available are extremely small. The main items are seasonal farm products or simple processed foods such as dried fruits, juices, vinegar, olive oil, wine, and some bakery products and pasta. One of the smallest industrial dairy enterprises has started with the production of fermented products in 2007 which are now widely available. Similarly, a poultry processing firm in 2010 offered organic meat and meat products in a major retail chain. However, there are still rather large challenges ahead for the Slovenian organic agro-food sector particularly to increase its market presence and assure stable supply of produce.

2.2 Slovenian organic food market volume

Only recently a rather comprehensive research project on Slovenian organic market development and domestic organic farms performance was carried out (Slabe et al., 2010). One of the central objectives of the study was to evaluate the organic food market volume with the evaluation of the sales channel structure and the share of the domestic products in total market supply. That part of the analysis was based on (i) in-depth interviews with the key market players which cover around 80-90% of total organic food turnover (ii) analysis of secondary information (mainly 2009 annual reports and other publicly available business documents) and (iii) a detailed survey of all national organic farmers markets including in-depth interviews with the market coordinators and a sample of approximately 30% of the farmers which were registered sellers on these markets in 2010.
The total estimated market value of organic food products and beverages in Slovenia for the year 2009 was 34.5 million EUR (Slabe et al., 2010). If this figure is compared to the total households expenditure for food and beverages in 2009 (SORS, 2011) the organic food and beverages represents approximately 1% of the budget. On average, per capita expenditure for organic foods and beverages respectively amounts to 17 EUR. If these two figures are compared with selected old EU countries this is relatively low. For example in Slovenia's

neighbour Austria per capita consumption was 104 EUR in 2009 with 6% organic share of food market. However, Austria is one of the leading countries in the world with regards to per capita consumption as well as share of the total food market. Higher per capita consumption and food expenditure share is found only in Denmark with 139 EUR and 7.2% respectively. If the data about the organic market are compared with the other new EU member states, Slovenia ranks the highest both in per capita consumption and share of the organic food in the total food market. Consumers in Estonia are the nearest by both indicators, since they spent 8.8 EUR on organic food which represents 1% of their total food spending. In Czech Republic per capita consumption was 7 EUR with 0.7% expenditure share, in Hungary 3 EUR and 0,3% respectively., whereas in Poland consumers spent on average 1.3 EUR on organic food annually, which is 0,1% of food budget. (Kilher et al., 2011).

The organic market in Slovenia has been growing at an annual rate between 10-15% the period between 2005 and 2009. The largest growth was in the category of fresh vegetables and fruits. The study revealed that the majority of the organic food and beverages sales are done through retail shop of different categories. It is estimated that more than 84% of Slovenian organic turnover is covered by these intermediaries. Direct selling on farm has 11% share while direct selling on the specialised farmers markets amounts to almost 5%. When the origin of organic products was estimated the study revealed that less than 5% of sales in retail channel is from Slovenia and respectively direct selling on farms or farmers markets is comprised of only domestic produce. The imports are mainly from the EU countries (Austria and Germany).

Sales channel	Market value *(in mio EUR)*	Channels share *(in %)*	Share of domestic *(in %)*
Retail shops *(conventional, discount, specialised)*	29.0	84.1	< 5
On-farm	3.8	11.0	100
Organic farmers' markets	1.7	4.9	100
Total	34.5	100	

Table 2. The structure of Slovenian organic food market by the sales channel, 2009. Source (Slabe et al., 2010)

Around 80% of the organic products of the domestic origin (both fresh and processed) was sold directly by the farmers. Approximately 2/3 of that value was created through the on-farm sales while the rest was realised through the organic farmers' markets. In the marketing season 2009-10, there were 13 organic farmers' markets in different cities of Slovenia operating once a week. The largest organic market was in Ljubljana with 26 vendors of which 2 were artisanal organic processors and the rest were farmers with fresh products. The organic farmers markets in other towns are somewhat smaller usually with between five to ten vendors. The organic market coordinators and the vending farmers perceived a growing interest of consumers in the last three to five years for the organic food. Also the interviews in the prevailing marketing channel for organic food and beverages in Slovenia, namely the retail shops, indicated a high level of consumers' demand for domestic organic products. This was supported by the observations at the outlets where many retailers explicitly indicated the organic products of domestic origin. The interviewed retailers also stated that they are interested in meeting consumers' demands for domestic organic products but are hampered by inadequate supply, both regarding the quantities and the range. Also low level of business professionalism for many producers was mentioned, in particular inappropriate preparation of the products, lack of reliability, absence of producers' organisations and sometimes also unrealistic price expectations. When the issue of low presence of domestic organic food in retail supply channel was presented to the organic farmers, they saw the main obstacles in their insufficient production capacity which is related mostly to the small farm size. Certainly, an important disincentive is also lower profitability of channelling the products through the intermediaries. Many farmers stated that direct selling is by far the most preferable marketing channel for organic food. The research has also identified several new forms of organic food sales within its duration (2008-2010), such as box schemes in both forms as direct and indirect sales. Furthermore, a institutions similar to Community Supported Agriculture was established. There has been also a significant increase of the number of registered enterprises involved in the trade of organic food products and also new local organic farmers' markets emerged. It can be assumed that the new selling channels and forms of direct sales is a response of innovative and proactive farmers to the consumers' demand. Contemporary food consumers require innovative selling methods with upgraded service. The farmers that are able to adequately answer to these needs will acquire premium market positions and sustainable competitiveness. For example in the case of box schemes, the farmer obliges himself to supply consumers with products in the agreed amount and frequency, while the consumers subscribe for the whole season and usually pay an agreed average yearly fee. The benefit is on both sides. The consumers have an assurance regarding the production methods and they normally pay less for the same amount of produce which is fresh and locally produced. The farmer, on the other side reduces risk of insufficient demand, but more importantly the actual marketing activity is minimised and the farmer can focus primarily on production.

2.3 Organic food price premiums

Organic food production is determined with lower yields when compared to conventional farming and therefore the price differential is an important purchasing barrier for majority of consumers. On the other side the price premium affects the organic farmers' profitability and therefore business attractiveness. There is only limited information available on prices for organic products in the EU (European Commission, 2005). However, as a general rule, organic products receive a higher price than conventional products, but prices diverge depending on the market and on the product.

The research on organic sector in Slovenia (Slabe et al., 2010) includes also a rather detailed analysis of price premiums for organic food products at the retail level in major marketing channels. The price scan was performed in two periods (June and October 2009) in order to reflect the difference of the season. In total 65 products pairs of organic and conventional categories were included and prices were scanned in four outlet types (farmers' market, specialised shops, conventional retailers, discount retailers). In total almost thousand entries were obtained through the price scan. The price premiums for organic products were calculated with the reference to the conventional counterpart. The organic price premiums

are the percent increase over conventional prices and are calculated by subtracting the conventional price from the organic price and dividing the difference by the conventional price. Than the price premiums were aggregated by the product groups and marketing channels by simple arithmetic mean.
The average price premium estimated for the organic food products in Slovenia for the year 2009 was 87% with large variations between product groups and distribution channels. On average the price premium for the aggregate "oils and fats" was 146%, following by "vegetables" (104%) and "fruits" (88%). For some of these products we are able to compare their individual price premiums with the EU15 estimates from the year 2002 (European Commission, 2005). Potato, for example reached 143% price premium in Slovenia, whereas in EU15 has ranged from 30% in Ireland to 170% in Greece. For apples the range of price premiums in EU15 was even more extreme from 37% in Sweden to 283% in Portugal, whereas in Slovenia it was 131% despite rather considerable domestic supply. Organic dairy products were priced on average at 73% higher than conventional substitutes, whereas on EU15 average the consumer price premium for organic milk was about 50%, with extremes in Greece (129%), Portugal (124%) and Italy (117%).

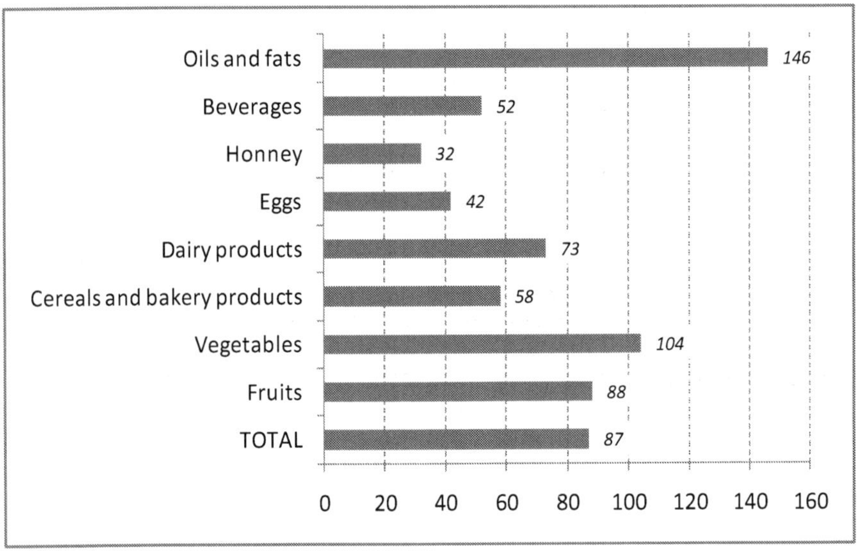

Fig. 1. Price premiums for organic food and beverages in Slovenia (2009)

For eggs the price premium was relatively low (42%), which is probably affected the fact that this sector is highly developed and efficient, but also proximity of the Austrian egg producers and their attempts to enter the Slovenian market. Price premium for eggs in Austria are at 25%, whereas in Denmark it is only 17% (European Commission, 2005). Again price premiums were high in Greece (231%), Spain (208%), Portugal (136%) and Italy (121%). The lowest price premium was found for organic honey which was on average 32% more expensive than its conventional substitute. Honey quality and hence the prices are relatively high in Slovenia and average consumers perceive conventional honey produced by small and middle size beekeepers as natural product, often labelled with the national quality mark. Therefore it is difficult to realise a higher price differential for organic honey.

It can be concluded that the price premiums for organic food were in general highest for fresh products (fruit and vegetables) and less for processed ones. Seemingly, consumers perceive guaranteed free from harmful substances in combination with fresh as the key attributes where they are willing to pay more and organic fruits and vegetables are therefore most appropriate food aggregates.

When the aggregated price premiums for organic food in Slovenia were compared across the marketing channels it was revealed the highest are in the conventional retail chains (136%), following by the discount retailers (104) and specialty shops (100%). The price premiums on the organic farmers' markets were found to be the lowest at 78%. In general the quality of agricultural produce sold at the Slovenian markets (not only organic ones) is considerably higher than the quality in supermarkets and consequently the prices are also higher. Therefore, it is somehow surprising that the price premiums for organic food on the markets were the lowest. Some explanation can be found in the fact that virtually all the organic fresh products in retail are sold pre-packed, which adds considerable costs to the final price, since neither conventional nor organic fresh products at the farmers' markets are packed. However, the relatively high price premiums and its large variation are confirming the fact that the Slovenian organic food market is in immature and it is expected with its further expansion and development the premiums will eventually diminish which will stimulate consumer demand, but will turn away marginal and inefficient producers.

3. Consumer study on the purchasing behaviour for organic and integrated fruits and vegetables

3.1 The research objectives and methodology

There are insufficient studies on organic consumers in Slovenia, therefore the presented research aimed at elucidation and quantification of the impact of various determinants influencing purchasing behaviour of organic and integrated fruit and vegetables consumers. A country-wide survey has been conducted on a representative sample of 1027 households. Beside the socio-demographic identification of the respondents, the main part of the questionnaire can be divided into the following sections:

- general dietary patterns and lifestyle determination;
- overall fruit and vegetable purchasing behaviour;
- household's fruit and vegetable self-sufficiency level;
- acquaintance, believes and perception of OIFV and
- purchase frequency of OIFV.

On the basis of the acquired data a consumer behaviour model of qualitative choice a (Pindyck and Rubinfeld, 1991) has been developed. Focus of the empirical scrutiny was given to identify determinants that influence purchase frequency of the OIFV. Methodology selection was directed by the ability for adequate incorporation of the ordinal nature of data describing purchasing behaviour of OIFV. The model that satisfactorily fulfils the criterion falls within the group of models of qualitative choice - more specifically the ordinal probit model (Pindyck and Rubinfeld, 1991).

Following Greene (1997), the ordered probit model can be specified, as built around a latent regression:

$$y_i^* = \beta' x_i + \varepsilon \tag{1}$$

where y_i^* is an unobserved frequency of quality fruits and vegetables purchase, β' is the vector of unknown parameters and x_i is a vector of explanatory variables (which may be continuous or discrete) denoting attributes influencing purchasing behaviour of respondent and ε is the independently and identically normally distributed error term. The ordered probit model tests the null hypothesis H_0: $\beta'_k = 0$ for every explanatory variable denoting that the independent variable k does not have an effect in explaining changes in the probability of y_i.

Results from our survey provide information on the respondents' purchase frequency of organic and integrated fruit and vegetable which is ordinal with five categories. While y_i^* is unobserved, respondents actually report their purchase decisions by selecting one of the five categories. Values for y_i are 0 through 4, where 0 represents no purchase of organic and integrated fruit and vegetables and 4 represents 2-3 weekly purchases of such produce.

The analysis builds from the following hypotheses:

i. consumers of OIFV in Slovenia are mostly influenced by the qualitative characteristics and not by the price premium;
ii. barrier to purchase (demotivator) for OIFV organic are price and availability;
iii. the main motivator to prefer integrated and organic fruit and vegetable to conventional is superior quality;
iv. important quality characteristics to consumers are nutritive value, freshness, flavour or taste and general appearance.

The results of ordered probit models were interpreted by using the partial change or marginal effects on the probability of ordinal outcome. Estimation of the empirical model was conducted by using the LIMDEP software (Greene, 1999). In doing so, the independent variables - other than the one being examined - were held constant at their mean values. The calculation of quasi-elasticities is based on the results of marginal effects, i.e. partial derivatives of the probability function *(Y)*. Like "standard" elasticity coefficients, quasi-elasticity coefficients can be interpreted as the percentage impact of a unit change of an explanatory variable on the probability of the observed outcome.

3.2 Results and discussion

Before turning to the results from the consumer choice model, this section starts with some general results about the perception of organic and integrated fruit and vegetables by Slovenian consumers grasped from the survey.

Despite a rather short period of organic and integrated production presence in Slovenia and no explicit marketing activities the survey results show that consumers' awareness is generally high. As expected the highest rate of recall has been achieved for the term "bio" that is an equivalent for "organic" in Slovenian language. As much as 94.4% of respondents have associated these expressions with fruit and vegetable. More than two thirds relate the phrase "ecological" with food, whereas only 38% of the respondents were acquainted with the term "integrated".

An open ended question was prepared to acquire basic associations of the respondents with the analysed categories of fruit and vegetables. Results show a rather high degree of responds homogeneity, since the three most frequent replies represent more than 80%. Associations are positive and generally indicate correct basic understanding. However, results show that the respondents do not distinguish among the organic and integrated

categories and perceive them as synonyms. That's why we have treated the frequency responses as for one product category only.
In total replies were categorised into 48 standardised answers and the highest frequency (37.1%) was attributed to association related to "healthy food". With 23.4% follows the category "free from harmful substances" where replies like: chemical free, pesticide free and alike were aggregated. Direct associations regarding the production practice rank third with 21.7% of replies. Surprisingly low share of replies was associated with "environment" (1.7%) and better quality (1.5%). The highest frequency among the wrong associations went to "low calories" (0.7%).
Dependent variable in the consumer choice model was formed on the basis of the question where respondents were asked to evaluate (self reporting) purchase frequency of organic and integrated fruit and vegetable (OIFV). Responses were coded in an ordinal scale as presented in the Table 3 below.

Purchase frequency	Number of observations	Frequency (%)	Cumulative frequency (%)	Dependent variable (PURCHAS)
Never	261	27.97	27.97	0
Less than once a month	225	24.12	52.09	1
1-3 times a month	162	17.36	69.45	2
Once a week	181	19.40	88.85	3
More than once a week	104	11.15	100.00	4
Total observations	933*	100.00		

** Non-responses or 'I don't know' responses were omitted from the analysis*

Table 3. Purchase frequency for organic and integrated fruit and vegetables

As suggested from the survey results Slovenian consumers assert rather a high purchase frequency for OIFV, since almost one third of them buy this category of produce at least once a week. Roughly the same share of respondents (28%) is non-buyers of OIFV and one quarter of them are sporadic buyers with purchasing fewer than once a month. About 17% of respondents buy this product few times a month.
Table 4 aggregates variables employed in the evaluated consumer choice model, which gives us some further insights to the topic. The finding that Slovenian consumers relate attributes of health as the main association with OIFV is confirmed with the highest rank from the likert test followed by the environmental attribute. Interesting finding comes from the question regarding the price of OIFV that ranks on last position. It seems Slovenes do not consider themselves as price sensitive.
When asked to compare general appearance and taste of OIFV with the "ordinary", produce respondents evaluate the appearance to be less likely better than taste. However, the "level of disagreement" among respondents (standard deviation) is also higher.
Perception regarding the availability of the studied categories of fruit and vegetables in the most frequently used retail shop is inclined towards the answer "insufficient"; however again high standard deviation is observed. On the other hand, dependency between the origin of food and quality has been reported as highly important. As might be expected from the general characteristic of Slovenian rural economy the level of households' self-supply with fruit and vegetables is rather high since only 37% of respondents purchase more than half of the total consumption.

		Total average (st. dev.)	Y = 0 average (st. dev.)	Y = 1 average (st. dev.)	Y = 2 average (st. dev.)	Y = 3 average (st. dev.)	Y = 4 average (st. dev.)
No. of observations		933	261	225	162	181	104
Logarithmic transformation of households' annual income	*INC_LN*	12.24 (0.52)	12.09 (0.52)	12.29 (0.54)	12.30 (0.52)	12.30 (0.48)	12.30 (0.47)
Environmental attribute OIFV prod. environment friendly (likert scale, 1-7)	*ENVIR*	5.84 (1.56)	5.85 (1.66)	5.80 (1.53)	5.88 (1.68)	5.80 (1.44)	5.98 (1.48)
Health attribute: OIFV considered healthier (likert scale, 1-7)	*HLTH*	6.07 (1.35)	5.88 (1.55)	6.00 (1.30)	6.25 (1.21)	6.15 (1.25)	6.22 (1.32)
Price attribute: OIFV are too expensive (likert scale, 1-7)	*PRICE*	5.68 (1.55)	5.77 (1.59)	5.68 (1.41)	5.62 (1.61)	5.61 (1.60)	5.72 (1.56)
Visual attractiveness of OIFV (likert scale, 1-7)	*VISUA L*	4.71 (1.62)	4.40 (1.68)	4.41 (1.48)	4.78 (1.62)	4.98 (1.54)	5.38 (1.59)
Taste of OIFV deemed better (likert scale, 1-7)	*TASTE*	5.35 (1.51)	5.17 (1.56)	5.05 (1.59)	5.46 (1.46)	5.56 (1.40)	5.83 (1.31)
Availability of OIFV by frequently used retailers (likert scale, 1-7)	*RETAIL*	3.99 (1.91)	3.52 (2.01)	3.62 (1.89)	4.12 (1.82)	4.36 (1.70)	4.85 (1.77)
Perceived linkages between origin and quality (likert scale, 1-7)	*ORIG_ Q*	5.92 (1.33)	5.99 (1.39)	5.74 (1.46)	5.82 (1.31)	6.04 (1.10)	6.10 (1.33)
More than 50% of F&V purchased (0=N; 1=Y)	*BUYER*	0.37 (0.48)	0.23 (0.42)	0.29 (0.45)	0.47 (0.50)	0.44 (0.50)	0.48 (0.50)
At least one meal daily cooked by themselves (0=N; 1=Y)	*COOK*	0.57 (0.50)	0.50 (0.50)	0.44 (0.50)	0.60 (0.49)	0.60 (0.49)	0.63 (0.48)
Residence in rural area (0=N; 1=Y)	*RURAL*	0.63 (0.48)	0.60 (0.49)	0.48 (0.50)	0.63 (0.48)	0.73 (0.45)	0.54 (0.50)

Table 4. Definition of explanatory variables and descriptive statistics of data used in the model

Table 5 presents the parameter estimates from the ordered probit model of consumer choice for organic and integrated fruit and vegetables in Slovenia.

Based on the results of a likelihood ratio test (Pindyck and Rubinfeld, 1991) the model is statistically significant at 99% or above. The results of the $\chi 2$ test reveal that the differences between the model coefficients are statistically significant. However, the value of the likelihood ratio index (LRI) goodness-of-fit coefficient (Greene, 1997) is rather low (0.049), which implies that the model explains only a part of the variance within the dataset. However, the main purpose of this empirical work was not to maximise probability function but to evaluate impact of some attitudinal variables on purchase frequency. Rather low LRI values were also expected due to the fact the survey dataset comprised only rather general determinants influencing households' purchasing behaviour. For higher degree of explanatory capacity of the model clearer definition of determinants is needed in the future work.

Explanatory variable	*Ordered probit of purchasing behaviour*		*Marginal effects for various outcomes*			
	Coefficient	*t-statistic*	*Prob(Y=0)*	*Prob(Y=1)*	*Prob(Y=2)*	*Prob(Y=3)*
Constant	-4.29	-4.20	1.306	0.406	-0.181	-0.773
Households' yearly income	0.30	3.84	-0.092	-0.029	0.013	0.054
Environmental concern: OIFV not harmful	-0.08	-2.40	0.025	0.008	-0.003	-0.015
Health concern: OIFV considered healthier	0.10	2.63	-0.030	-0.009	0.004	0.018
Price consciousness: OIFV too expensive	-0.02	-0.67	0.006	0.002	-0.001	-0.003
Visual attractiveness of OIFV	0.10	3.82	-0.031	-0.010	0.004	0.018
Taste of OIFV deemed better	0.09	2.88	-0.026	-0.008	0.004	0.015
Availability of OIFV by frequently used retailers	0.13	6.00	-0.038	-0.012	0.005	0.023
Perceived linkages between origin and quality	-0.01	-0.05	0.001	0.000	-0.000	-0.000
More than 50% of F&V purchased	0.14	1.67	-0.043	-0.013	0.006	0.025
At least one meal daily cooked by themselves	-0.09	-1.08	0.028	0.008	-0.004	-0.016
Residence in rural area	-0.18	-2.11	0.056	0.017	-0.008	-0.033
Log likelihood function	-1155.27	/	/	/	/	/
Restricted log likelihood	-1215.00	/	/	/	/	
LR test $\chi 2$ (d. freedom.)	119.49 (11)	/	/	/	/	/
LRI	0.049	/	/	/	/	/

Table 5. Results of the consumer choice model

Results show that the income status of consumers considerably determines purchasing frequency for organic and integrated fruit and vegetable. As has been expected, purchasing frequency significantly increases with higher household disposable incomes; however the estimated marginal effects reveal non-linear patterns for this variable. A high quasi-elasticity coefficient for non-buyers (Y=0) ranking to 0.66 suggests that low income level very likely determines no purchasing of organic and integrated fruit and vegetables. Also the marginal effects revealed for this variable clearly suggest that higher frequency of purchase is closely related with households' disposable incomes (and vice versa). The corresponding quasi-elasticity for a consumer group that purchase such produce on a regular basis (once a week or more) is considerably lower (0.39), and therefore the intensity of the relation is smaller.

The model results with respect to the stated environmental concern reveal that respondents not considering production of organic or integrated fruit and vegetable as environment friendly (or they are indifferent to environmental aspects of production) are more likely to be among non-buyers. It however has to be further noted that the impact of environmental concerns on purchasing behaviour is significant, but not explicit. However, the highest relation is found for the non-buyer group (0.08). Rather inconclusive results regarding environmental concern might be further explained with low association between organic and integrated production practices and implications on environment by the Slovenian consumers.

If a respondent perceives organic and integrated fruit and vegetables being healthier than conventional products the probability (and frequency) of actual purchase is significantly higher. On the contrary, persons not considering quality products as healthier are more likely to be among non-buyers (quasi elasticity 0.10).

According to the model results, price consciousness has no significant impact on purchase of quality fruits and vegetables. Interestingly, non significant coefficient suggests that price of higher quality products is not a decisive element of purchasing behaviour. Purchasers are likely to continue buying such products notwithstanding higher prices. Surely, these results should not be considered a basis for an ultimate conclusion about low consumer price sensitivity for organic and integrated fruit and vegetables in Slovenia. Some additional and more sophisticated measuring approaches should be employed to confirm these indications.

However, it is confirmed by the model results that consumers consider the visual attractiveness (appeal) of fruits and vegetables when they make purchasing decision. Consumers which believe that visual appealing of organic or integrated fruit and vegetables is not satisfying (worst than conventional), are less likely to buy these categories of produce. The highest quasi-elasticity is 0.02, which is again linked with decision not to purchase and therefore, and for that group of consumers the impact of visual attractiveness on purchase frequency is rather low.

Taste appears to significantly affect the consumer preferences to purchase fruit and vegetables from organic or integrated production systems. Model results show that consumers perceiving these categories of fruits and vegetables as having superior taste comparing the conventional ones are more likely to be among buyers.

It can be further examined, however that the consumers do not relate quality of fruits and vegetables with their micro-origin. The coefficient estimating this determinant is insignificant and therefore based on this results potentials for "local supply" marketing strategy, turned to be less appropriate.

Market for organic fruit and vegetable in Slovenia might still be considered as insufficiently developed. Situation for integrated produce is slightly better; however awareness of consumers is very low for this category. Therefore it has been expected, that availability of such products at "my retailer" has significant role on the consumer purchasing behaviour. Model results clearly confirm these expectations and favour strong emphasis on distribution strategy.

The level of self-sufficiency showed to be rather high for Slovenian households; however, result from the model doesn't confirm the expected inverse relations. The fact that households buy more than a half of fruits and vegetables does not have a statistically significant effect on frequency of purchase of organic and integrated fruit and vegetable.
Similar holds for dietary habits of households. Results from the model confirm that meal preparation is not significantly related with higher probability to purchase organic or integrated fruit and vegetables. Consumers which prepare at least one main meal within the household a day might be named as "traditional eaters" and they are not necessary the main purchasers of organic and integrated fruit and vegetable. Reversely; marginal effects suggest that for this group of respondents the probability for frequent purchase (outcome Y=3) decrease, however with low quasi elasticity.
Results for the last variable suggest that consumers from rural areas are generally less likely to buy organic and integrated fruit and vegetables. This might be related either to the problem of availability for these produces in rural area, insufficient awareness of consumers but partially also to the household self-supply with fruit and vegetable.

4. Conclusion

The organic food market in Slovenia started to develop sporadically immediately at the beginning of the period of economic transition, whereas the adequate certification system evolved several years later with first farmers being certified before the year 2000. Since then the both sides of the market are on rather sharp increase however, the supply side has some deficiencies. Namely, the area under the organic farming is increasing and number of farmers likewise, but is this is not sufficient to satisfy demand. Mainly due to the phenomenon of passive organic farmers, especially within the animal husbandry sector that are not marketing their produce into the organic sales channels. The organic farming area growth is actually mainly within the pastures and grassland categories and the farmers' main motivation to enter the organic farming is additional budgetary transfer without actual intention to be active in marketing organic products. Therefore, domestic production is relatively stagnant and Slovenian consumers can thus hardly obtain organic products of Slovenian origin in the massive food distribution channels, where the growing demand is mainly satisfied with imported products. It is therefore understandable that the involved organic food consumers prefer to purchase directly from the farmers, even if this is related with higher transaction costs, such as driving to the farm or visiting the farmers' market, etc. However, beside the assurance that the purchased products are fresh and local, also the prices and price premiums are lower when purchasing directly. In terms of per-capita spending on organic food and share of organic food expenditure in total food budged Slovenia with 17 EUR and 1% respectively ranks in the upper middle stratum of the EU27, where all new member states shows much lower values. This might indicate that the Slovenian organic food market is in the progressive stage of development however the results from the price scan and margin evaluation prove the opposite. Price premium for organic products in Slovenia are comparatively high (87%) with great variation between the product categories, where the fresh products (particularly fruits and vegetables) tends to have highest price margins. It is worth to notice, that when the price margins were analysed by the sales channels the lowest figures were found for direct selling either on farm or farmers' markets, whereas the margins were by far the highest in the conventional retail chains. This somehow controversial with the fact that only about 15% of the total Slovenian organic food turnover is realised through direct selling, the remaining share goes to different formats of retail.

An interesting conclusion regarding the attitudes and perception of organic and integrated fruit and vegetable among Slovenian consumers is that they do not associate these categories with environmental dimension. The most frequent association in this respect is health ("organic or integrated is healthy") and free from harmful substances, whereas less than two percent of respondents relate to positive impacts on environment. Implication from this finding is that much wider promotional and educational activities regarding organic and integrated fruit and vegetable is needed in Slovenia. Especially awareness for integrated production is very low (only 38% of respondents relate this term with food) despite the fact that the actual production of such fruit in Slovenia is considerable high especially for staple fruit such as apples and pears.

When the factors that might affect frequency of organic and integrated fruit and vegetable consumption in Slovenia were evaluated we firstly found that almost one third of responded reported they buy this category of produce at least once a week. However, roughly the same share of respondents (28%) falls within the group of non-buyers. Surely, using the self reporting method for purchase frequency of a sub-group of product that is not clearly defined (or homogeneously perceived) might lead to over evaluation. However, very low share of incorrect associations (e.g. low calories) and high frequency of non-specific association (healthy and free of harmful ingredients) might indicate that consumers attribute "organic or integrated" category uncritically. They "believe" they purchase this category of fruit or vegetable if the produce fulfils some general stereotypes e.g. produced within extensive orchard. This conclusion again supports already identified need for more effective promotion and consumer education.

The ordinal probit model was constructed using purchase frequency for organic and integrated fruit and vegetables as dependent variable and 11 selected qualitative determinants as independent variables. Results of the model are statistically significant whereas the goodness-of-fit indicator (LRI) is rather low. However; at this stage of the research even with low explanatory capabilities for variance, the model gives valuable insight into the organic and integrated fruit and vegetable purchasing behaviour of Slovenian population.

The most significant impact on purchase frequency has availability of organic and integrated fruit and vegetables in the shop where respondents make majority of their shopping. Clearly, the importance of product availability favours effective distribution activities. As it can be observed on the Slovenian market, the emerging trends of direct purchase (e.g. farmers' market, on farm buying, box schemes) and attempts of the retail chains to explicitly communicate their local fruit and vegetables sourcing (ethnocentrism, and/or low food miles) confirm this conclusion.

Model confirms important inverse implication of disposable household income on purchase frequency, where beside the affordability effect (these category of produce is more expensive) also education and awareness might influence the result. The two basic criteria of quality - taste and visual attractiveness have both significant effects on the frequency of purchase. If consumers perceive organic and integrated fruit and vegetables as superior in terms of taste and visual appeal, probability of a more frequent purchase is higher (it has however to be noted that quasi elasticity for both determinants is low). Nevertheless, descriptive results of the survey suggest that consumers claim they are often prepared to "sacrifice" superior visual attractiveness for the organic and integrated category, but the taste should be better. This might be a useful guideline for business development strategies. Relation between environmental concern and organic or integrated fruit and vegetable is significant but rather inconclusive. Marginal effect are positive for non buyers, whereas for frequent buyers are negative. These results are accordant with conclusion about low association between environment and organic and integrated production. Insignificant relation has been evaluated for meal preparation patterns, however they are rather explanatory. Probability of being frequent buyer of organic or integrated fruit and vegetable increases with the fact that a respondent is only sporadically cooks meals at home. Traditional eating patterns are not a characteristic of aware and affluent consumers. They do increase the share of food consumed away from home and are also disposed to modern food categories where organic and integrated produce surely can be classified.

The contribution adds towards a better understanding of demand for organic and integrated fruit and vegetable in emerging markets. Despite this, additional research would be needed to understand sufficiently such a complex processes as food choice. In this respect, it would be interesting to study in greater detail consumers' attitudes by different distribution channels and strategies for organic and integrated fruit and vegetables. This would help to prioritise specific attributes and to evaluate price sensitivity of consumers for organic and integrated produce.

5. References

Achilleas, K., & Anastasios, S. (2008). Marketing aspects of quality assurance systems: The organic food sector case. *British Food Journal,* 110 (8), pp. 829-839; ISSN: 0007-070X.

Batte, M.T., Hooker, N.H., Haab, T.C. & Beaverson, J. (2007). Putting their money where their mouths are: consumer willingness to pay for multi-ingredient, processed organic food products. *Food Policy,* 32 (2), pp. 145-159, ISSN: 0306-9192.

Bellows A. C., Onyango, B., Diamond, A. & Hallman W.K. (2008). Understanding consumer interest in organics: production values vs. purchasing behavior. *Journal of agricultural and food industrial organization.* 6 (1), pp. 1-28, ISSN: 1542-0485.

Boccaletti, S. & Nardella, M. (2000). Consumer's willing to pay for pesticide-free fruit and vegetable in Italy. *International Food and Agribusiness Management Review,* (3), pp. 297-310, (ISSN: 1559-2448.

Botonaki, A., Polymeros, K., Tsakiridou, E. & Mattas, K. (2006). The role of food quality certification on consumers' food choices adequate marketing strategy for the effective promotion of certified food products. *British Food Journal,* 108 (2), pp. 77-90, ISSN: 0007-070X.

Canavari, M., Nocella, G. & Scarpa, R. (2005). Stated willingness-to pay for organic fruit and pesticide ban-an evaluation using both web-based and face-to face interviewing. *Journal of Food Products Marketing,* 11(3), pp. 107-134, ISSN 1045-4446.

Connor, J.M. (1994). Northern America as a precursor of changes in Western European food purchasing patterns. *European Review of Agricultural Economics,* 11 (2), pp. 155-173, ISSN 1464-3618.

European Commission (2005) *Organic farming in the European Union. Facts and figures.* European commission, DG Agriculture and rural development. Available at:

http://ec.europa.eu/agriculture/organic/files/eu-policy/data-statistics/facts_ en. pdf [Accessed 10.07.2011]

Eurostat. (2007). *Organic farming in the EU*. Eurostat News Release 80/2007, Statistical Office of The European Communities, ISSN 1977-0316.

Halberg, N., Alroe, H., Knudsen, M. T. & Kristensen, E. S. eds., (2006). *Global development of organic agriculture: Challenges and promises*. CABI Publishing, ISBN-10: 1845930789.

Govindasamy, R. & Italia, J. (1999). Predicting willingness-to-pay a premium for organically grown fresh produce. *Journal of Food Distribution Research*, 30, pp. 44-53, ASIN: B00063FLIW.

Greene, W.H. (1997). *Econometric analysis*. 3rd ed. New Jersey, London: Prentice Hall International, ISBN O-471-53233-9.

Greene W.H. (1999). *Limdep version 7.0: User's manual*. New York: Econometric Software Inc.

Grunert, S.C. & Juhl, H.J. (1995). Values, environmental attitudes and buying of organic foods. *Journal of Economic Psychology*, 16(1), pp.39-62, ISSN: 0167-4870.

Hamm, U., Gronefeld, F., & Halpin, D. (2002). *Analysis of the European market for organic food - Organic Marketing Initiatives and Rural Development (OMIaRD) Volume 1*. Aberystwyth: University of Wales, 157 p, ISBN 0-9543279-4-7.

Kihlberg, I. & Risvik, E. (2007). Consumers of organic foods – value segments and liking of bread. *Food quality and preference*. 18, pp. 471-481, ISSN: 0950-3293.

Kilcher, L., Willer, H., Huber, B., Frieden, C., Schmutz, R. & Schmid, O. (2011): *The Organic Market in Europe*: 3rd edition May 2011, SIPPO, Zurich and FiBL,Frick, 184 p, ISBN 978-3-03736-186-3.

Krystallis, A. & Chryssohoidis, G. (2005). Consumers' willingness to pay for organic food: Factors that affect it and variation per organic product type. *British Food Journal*, 107 (5), pp. 320 – 343, ISSN: 0007-070X.

Lambert N. (2001). Food choice, photochemicals and cancer prevention. In Lynn, F. J., Einar, R., & Schifferstein, H., eds. *Food, people and society. A European perspective of consumers' food choice*. Springer –Verlag, pp. 131-151, ISBN-10: 3540415211.

Loureiro, M.L., McCluskey, J.J. & Mittelhammer, R.C. (2001). Assessing consumer preferences for organic, eco-labeled, and regular apples. *Journal of agricultural and resource economics*, 26 (2), pp. 404-416, ISSN 0162-1912.

Latacz-Lohmann, U. & Foster, C. (1997). From Niche to mainstream strategies for the marketing of organic food in Germany and the UK. *British Food Journal*, 99 (8), pp. 275-282, ISSN: 0007-070X.

MAFF. (2006). *Akcijski načrt razvoja ekološkega kmetijstva v Sloveniji do leta 2015*. Ljubljana: MAFF (Ministry of the Republic of Slovenia of Agriculture, Forestry and Food), 72 p, ISBN 961-6299-73-5.

MAFF. (2011). *Analiza stanja ekološkega kmetijstva v Sloveniji*. Ljubljana: MAFF (Ministry of the Republic of Slovenia of Agriculture, Forestry and Food). Available at: http://www.mkgp.gov.si/si/o_ministrstvu/direktorati/direktorat_za_kmetijstvo /starasektor_za_sonaravno_kmetijstvo/oddelek_za_kmetijstvo_in_okolje/kmetijsk o_okoljska_placila/ekolosko_kmetovanje/ekolosko_kmetijstvo_dejstva_in_podatki /7_analiza_stanja_ekoloskega_kmetijstva_v_sloveniji/ [Accessed 10.07.2011].

de Magistris, T. & Gracia, A. (2008). The decision to buy organic food products in Southern Italy. *British Food Journal*, 110 (9), pp. 929-947, ISSN: 0007-070X.

OECD. (2003). *Organic Agriculture: Sustainability, Markets and Policies.* Paris: OECD (Organisation for Economic Cooperation and Development), 408 p, ISBN: 9789264101517.

Pindyck, R.S. & Rubinfeld, D.L. (1991). *Econometric models and economic forecasts* (3rd international edition). New York: McGraw-Hill Inc, 436 p, ISBN-10: 0079132928.

Raynolds, L. T. (2004). The Globalization of organic agro-food networks. *World Development.* 32(5), pp. 725-743, ISSN: 0305-750X.

Sahota, A. (2008). The Global Market for Organic Food and Drink. In: Willer, H., Yussefi-Menzler, M., & Sorensen, N. eds. *The world of organic agriculture: Statistics and emerging trends 2008.* Bonn: IFOAM and Frick: FiBL, pp. 53-57, ISBN 978-3-934055-99-5.

Schifferstein, H.N.J. & Oude Ophuis, P.A.M. (1998). Health-related determinants of organic food consumption in The Netherlands. *Food Quality and Preference,* 9 (3), pp. 119-133, ISSN: 0950-3293.

Slabe, A. (2002). *Organic Farming in Slovenia.* Frick: FiBL. Available at: http://www.organic-europe.net/country_reports/ slovenia/ default.asp [Accessed 19.12.2008].

Slabe, A., Kuhar, A., Juvančič, L., Tratar-Supan, A.-L., Lampič, B., Pohar, J., Gorečan, M., Kodelja, U. (2010). Analysis of the status and potentials for the growth of organic products in the light of achieving the goals of the APOF. Final report, Proj. no. V4-0514-09.

Torjusen, H., Sangstad, L., O'Doherty-Jensen, K., & Kjærnes, U. (2004). *European consumers' conceptions of organic food: A review of available research.* Oslo, National institute for consumer research, 150 p, ISBN 82-7063-394-1.

Thøgersen, J. & Ölander, F. (2002). Human values and the emergence of a sustainable consumption pattern: A panel study. *Journal of Economic Psychology,* 23(5), pp. 605-603, ISSN: 0167-4870.

Thompson G.D. & Kidwell J. (1998). Explaining the choice of organic produce: cosmetic defects, prices, and consumer preferences. *American Journal of Agricultural Economics,* 80, pp. 277-287, ISSN 1467-8276.

Thompson, G.D. (1998). Consumer demand for organic foods: what we know and what we need to know. *American Journal of Agricultural Economics,* 80, pp. 1113-1118, , ISSN 1467-8276..

Underhill S.E. & Figueroa E.E. (1996). Consumer preferences for non-conventionally grown produce. *Journal of Food Distribution Research,* 27, pp. 56-66.

Viaene, J., Verbeke, W. & Gellynck, X. (2000). Quality perception of vegetables by Belgian consumers. *Acta Horticulturae.* 524,pp.89-96, ISSN 0567-7572.

Vermeir I. & Verbeke, W. (2006). Sustainable Food Consumption: Exploring the Consumer "Attitude - Behavioral Intention" Gap. *Journal of Agricultural and Environmental Ethics,* 19, pp. 169-194, ISSN: 1573-322X.

Weaver R.D., Evans D.J. & Luloff A.E. (1992). Pesticide use in tomato production: consumer concerns and willingness-to-pay. *Agribusiness,* 8, pp. 131-142, ISSN: 1520-6297.

Wier M. & Calverley C. (2002). Market potential for organic foods in Europe. *British Food Journal,* 104 (1), pp. 45-62, ISSN: 0007-070X.

Wier, M., Hansen, L.G., Andersen, L.M. & Millock, K. (2003). Consumer preferences for organic foods. In: *Organic agriculture: Sustainability, markets and policies.* CABI Publishing. pp. 257-271, ISBN: 9789264101517.

Willer, H., Yussefi, M. & Sorensen, N. Eds. (2008). *The World of Organic Agriculture - Statistics and Emerging Trends.* IFOAM, ISBN 978-3-934055-99-5.

Yiridoe, E.K., Bonti-Ankomah, S. & Martin, R.C. (2005). Comparison of consumer's perception towards organic versus conventionally produced foods: a review and update of the literature. *Renewable Agriculture and Food System,* 20(4), pp. 193-205, ISSN: 1742-1705.

3

Should I Buy Organic Food? A Psychological Perspective on Purchase Decisions

Christian A. Klöckner
Norwegian University of Science and Technology
Norway

1. Introduction

The individual consumer - although placed at the end of the production chain - plays an important role in establishing and developing the market for organic food. It is the final purchase in a supermarket, in a health-food shop or on a farmers' market that creates the demand that eventually sustains organic agriculture. Purchasing food is by no means a simple decision. It can be split into a series of interlaced decisions such as: When do I do my food shopping (e.g., after work, on a Saturday, under time pressure or not)? Where do I go (e.g., local supermarket, hypermarket, health-food store, farmers' market)? How much money do I want to or can I afford to spend? Which classes of products do I want to purchase? Within each class: what is the specific produce I purchase? Decisions made earlier in this chain impact the context of decisions made later. If for example a decision for shopping in a supermarket instead of a health-food store is made the variety of produces is different which impacts the produces that are taken into consideration. If food shopping is done under time pressure, time invested to make decisions is dramatically reduced and mental shortcuts or routines take control. Furthermore, the decision process might be non-linear, jumping back and forth between some of the aforementioned levels.

Psychological research has produced a large number of studies that allow insight into the complexities of this decision making process. It has been shown that consumers' purchase decisions at a given point in time and in a specific context are determined by a variety of psychological and contextual factors and their interactions. Some of them will be reviewed in this chapter. Based on previous research the following aspects will be discussed: How do values, attitudes and concerns for health or the environment impact the purchase of organic food? How do visibility, availability and perception of prices contribute? What is the role of trust? How can environmental and health psychological models contribute to understanding organic food purchase? How are organic food labels perceived and used in decision making? Finally, an integrated framework model will be suggested in the last section before drawing conclusions for future research.

2. General motives to buy organic food: Values, concerns and attitudes

One tradition in psychological research on the purchase of organic food produce focuses on identifying general motivations that may lead to favouring organic agriculture and eventually preferring the organic over the conventionally produced alternative when making a decision. This research tradition has its roots in value and attitude psychology and assumes that value orientations and attitudes are important determinants of people's behaviour. Before analysing their impact on the purchase of organic food in more detail, the three core concepts of this section shall be defined and distinguished from each other in the first paragraph of each subsection.

2.1 Values

One of the most basic psychological concepts is a *value*. Schwartz (1994) defines values as *"desirable transsituational goals, varying in importance, that serve as guiding principles in the life of a person or other social entity* (page 21)". This definition outlines four important features that characterize values: (a) they define what is morally desirable to achieve for a person, (b) they are allocated on a very general level which makes them applicable across situation, (c) they may vary in importance between different cultures, people or situations, and (d) they motivate behaviour because they guide goal-setting and choice of action. Schwartz (1992) furthermore suggested a categorization of ten basic value orientations (power, achievement, hedonism, stimulation, self-direction, universalism, benevolence, conformity, tradition, security) which has been widely adapted in cross-national studies as well as in various behavioural domains.

Grunert and Juhl (1995) applied the Schwartz value inventory in a study on Danish school teachers to determine the relation between basic value orientations, general environmental attitudes and organic food consumption. They were able to show that value orientations that fall into universalism were most characterizing for what they called "teachers with green attitudes", but also self-direction, stimulation and hedonism to a smaller degree. In a second step they demonstrated that "green" teachers much more likely occasional or regular buyers of organic food. Dreezens et al. (2005) used Schwartz' value system to analyse the relation between beliefs about organic food, attitudes towards organic food and basic value orientations. They found a positive relation between positive attitudes towards organic food and universalism and a negative with power. Furthermore, they could show that this relation is only indirect, mediated by beliefs about organic food (e.g., agreeing that organic food is good for the environment, tastes better, is healthier, etc.). The effects were of a moderate size. In a similar survey conducted with a population sample in Australia Lea and Worsley (2005) found that self-transcendence values - especially personally valuing nature, the environment and equality - were positively related to holding positive beliefs about organic food. However, the relation found was fairly weak. In a Norwegian survey Honkanen et al. (2006) found on the other hand a rather strong relation between the ecological shade of ethical food choice motives and a positive attitude towards organic food which eventually impacted the intention to buy organic food positively. Weak or no relations were found between political motives or religious motives and pro-organic attitudes.

In a qualitative study Makatouni (2002) analysed the value orientations that were relevant for preferring a variety of organic produce over their conventionally produced counterparts in a sample of British parents of 4-12 year old children. The most relevant value embraced was preserving health of themselves and their families, but also protecting the environment and animal welfare were values important to people that preferred organic food alternatives. Health protection would fall under the security value in the Schwartz system, animal welfare and protection of the environment would in Schwartz' understanding be part of a universalistic value orientation. What is interesting here is that preference for organic food can be attributed to two very different, almost opposed value orientations. In a similar approach Baker et al. (2004) compared a sample of German with a sample of UK citizens in another qualitative study and identified health/enjoyment, belief in nature, and animal welfare as the most prominent value orientations driving organic food consumption in Germany, whereas in the UK health/enjoyment/achievement and respect for others/workers emerged as the dominant value orientations, interestingly omitting nature totally. Again the interesting finding is that organic food consumption can have motivations that stem from very different basic value orientations in the Schwartz system.

Two conclusions can be drawn from the analysis of the relations between values and organic food choice: (a) the relation is usually rather weak and indirect, mediated by other variables such as beliefs and attitudes, (b) different, sometimes even opposing value orientations are potentially motivating organic food choice. Some people prefer organic food because they value their health and believe in positive health effects of organic food (value dimension: security), some people prefer organic food, because they want to protect nature, animals, or workers (value dimension: universalism), some prefer organic food because of hedonistic motives (e.g., better taste).

2.2 Attitudes

Eagly and Chaiken (1993) define *attitudes* as *"a psychological tendency that is expressed by evaluating a particular entity with some degree of favour or disfavour* (page 1)". This definition names three key features of an attitude: (a) it is linked to an entity (an object, a person or a behaviour), (b) it includes a general evaluation of this entity as desirable or not, and (c) is a psychological predisposition that might or might not be expressed in certain behaviours. Fishbein and Ajzen (1975) conceive of attitudes as the general summation of all activated beliefs about the attitude object, with beliefs being the likelihood of a certain outcome of a course of action times its evaluation. Attitudes are in contrast to values connected to specific objects and therefor much less general and transsituational.

Already in the previous section attitudes were introduced as potential mediators between very general value orientations and consumption of organic food or at least the intention to do that. A lot of papers have analysed the relation between attitudes and purchase of organic food, the most interesting are outlining what the most important positive and negative beliefs about organic food and conventional alternatives are that constitute the attitude. A short summary will be given in the remainder of this section. Storstad and Bjørkhaug (2003) analysed attitudes among farmers and consumers in Norway and found that attitudes consisting of pro-environmental beliefs were the only psychological variable positively influencing the purchase of organic food. Pro-animal welfare attitudes were not important, basically because in the case of Norway also conventional agriculture has the image of being animal friendly (Nygård & Storstad, 1998). In a study with inhabitants on a small Scottish island Michaelidou and Hassan (2008) were able to show that the link between positive attitudes towards organic food and the intention to buy it was strong. Furthermore, they found that concerns for food safety, health consciousness and an ethical self-identity were components that significantly contributed to this attitude. Based on data from a national survey Onyango et al. (2007) identified the following food attributes as the most important components of a pro-organic food attitude in the US: (a) naturalness, (b) vegetarian-vegan, (c) production location, (d) familiarity (negative impact). De Magistris and Gracia (2008) conclude in a study based on a survey in southern Italy that beliefs about positive outcomes for health and environment are the most important facets of a pro-organic food attitude and that this attitude is positively impacted by available information on the organic food market.

The aforementioned studies are just a small sample of the available literature on attitudes towards organic food. What becomes clear is that pro-organic attitudes are multi-faceted and many different beliefs contribute. Building on a literature review published by Hughner et al. (2007) figure 1 summarizes the most important beliefs that have been connected to organic food in previous studies and that contribute to forming an attitude towards organic food (e.g., Hughner et al., 2007; Storstad & Bjørkhaug, 2003; Michaelidou & Hassan, 2008; Onyango et al., 2007; De Magistris & Gracia, 2008; Schifferstein & Ophuis, 1998; Özcelik & Ucar, 2008; Padel & Foster, 2005).

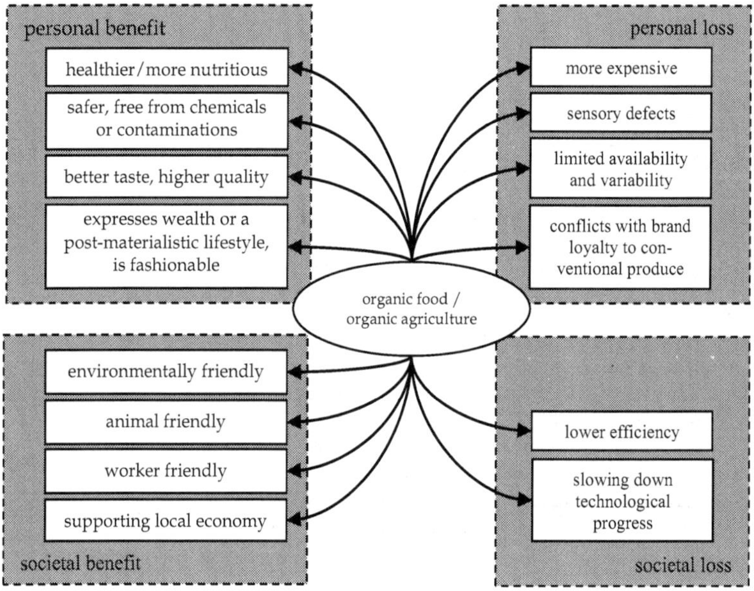

Fig. 1. Beliefs connected to organic food

In figure 1 the beliefs that have been previously found have been arranged into four groups: (a) personal benefits a person ascribes to the consumption of organic food, (b) societal benefits the person ascribes to organic agriculture, (c) personal losses that a person ascribes to the consumption of organic food, and (d) societal losses associated with organic agriculture. All of those aspects are subjective beliefs, which means it is irrelevant if the assumptions underlying them can be supported by scientific findings or not. Usually, anticipated personal benefits and losses are more relevant for behavioural choice than societal benefits and losses, but a strong universalist value orientation (see section 2.1) makes societal benefits more salient. Furthermore, strongly felt moral obligations to protect the environment have been shown to reduce the importance of negative beliefs like the ones subsumed under "personal losses" in figure 1 (Klöckner & Ohms, 2009).

Although attitudes have been repeatedly shown to be a relevant predictor of organic food purchase a significant gap between pro-organic attitudes and actual consumption of organic food remains. This attitude-behaviour gap that has been described in many behavioural domains has been attributed both to variables mediating between attitudes and behaviour and variables moderating this relation (Armitage & Christian, 2003). The first line of reasoning refers to that attitudes may not be a direct predictor of purchase behaviour but only have an indirect effect via other more proximal variables. The second line of reasoning refers to the assumption that other variables might impact the strength of the relation between attitudes and behaviour. Both aspects lead to more complex models that will be debated in section 5.

2.3 Concern

Finally, *concern* is a less clearly defined construct. In the medical and health context concern has been described as a worry expressed by a patient or a strong negative emotion (Schofield, Green, & Creed, 2008). With respect to other people concern has also been understood as expressing sympathy and compassion for less fortunate others (Fox, 2006), a concept that might also be generalized to non-human creatures or the environment in general. What characterizes concern is therefore an emotional reaction to anticipated negative effects either for oneself or for other people which potentially leads to tendencies to act against the negative impact.

Although concern for health, food safety, the environment, animal welfare or agricultural workers have been discussed in section 2.2 already and although there is a certain overlap it makes sense to look into concern for health and food safety as these two motivators of purchasing organic food might function differently from the others discussed before. Health concerns connected to conventional food are the most relevant motivator to buy organic food (Hughner et al., 2003). Magnusson et al. (2003) found that health concerns are more important than environmental concerns. Padel and Foster (2005) outline that this is especially the case for people with children. Specifically, the absence of chemicals like artificial fertilizers or pesticides, growth hormones or antibiotics etc. in organic agriculture and thereby avoiding possible negative health effects have been named as motivators (e.g., Schifferstein & Ophuis, 1998; Ott, 1990). Furthermore, that organic food is free from genetic modification is another motivator with a connection to health concern (Baker et al., 2004; Makatouni, 2002). However, Verdurme et al. (2002) were able to show that not all people who purchase organic food are opposing genetically modified food. Makatouni (2002) also found that fear of animal diseases or food scandals associated with the conventional food industry may have an impact. What makes separating health and food safety concerns, which is an emotional reaction to a perceived health threat, from the other beliefs about organic food attractive is that this opens for applying health psychological models to the purchase of organic food. This will be pursued further in section 6.

3. Situational impacts and their subjective representation: Availability, visibility, and price

It is not surprising that situational conditions like availability, visibility or the price of organic food relative to conventional alternatives has an impact on purchase decisions. In this section their impact will be analysed in more detail, with a special focus on their subjective representation, because objective accessibility or price differences are not necessarily in accordance with subjective representations people have of them - and it are the latter that impact the decision.

3.1 Availability

In their overview paper Hughner et al. (2007) identified perceived lack of availability of organic food and inconvenience associated with the purchase process as one of the main barriers to organic food purchase. In a qualitative study with Italian customers by Zanoli and Naspetti (2002), people associated organic products as difficult to find. Padel and Foster (2005) found similar results in a UK sample and concluded that people reacted negatively to limited choice options (compared to conventional alternatives) and higher effort that needs to be put into buying organic food (e.g., additionally entering a health food store). In an unpublished interview pilot-study with Norwegian customers, limited accessibility of a full range of products in the organic food sector was named as the main barrier (Klöckner, 2008). In an analysis conducted with Turkish customers, availability of organic products was a better predictor of purchase frequency than anticipated environmental benefits (Ergin & Ozsacmaci, 2011). The most important predictor was trust (see section 4), followed by health considerations, availability and environmental benefits.

In a comparative review of organic food consumption in different European countries Thøgersen (2010) presents evidence for that the percentage of organic food consumption in a country is a function of influences from four different domain: political regulations (laws & subsidies), politically motivated marked development (certification, labelling, information campaigns), the demand side in the market (values, environmental concern, food culture, income level, etc.) and the supply side in the market. This last factor clearly reflects that in order to sustain a functioning organic food market opportunities for the customer have to be created and convenient distribution channels have to be used to make organic products available at the point in time and space where the food purchase decision is made. Not coincidentally sales increased in many countries substantially after the big supermarket chains entered the organic food market (Aschemann et al., 2007), which is most likely the combined outcome of increased availability and marketing activities. Thøgersen (2010) argues that consumers' attitudes, values and norms and the like are only relevant for a purchase decision within a decisional space defined by the opportunities the supply side creates, which makes availability and easy access to one of the key features in increasing organic food consumption. Very few customers are willing to go the extra mile to buy an organic product.

3.2 Visibility and shelf-placement

The impact of visibility, placement on the shelf and shelf space of organic produce compared to their conventional alternatives may be regarded as a sub-phenomenon of the aforementioned availability discussion, but analysing their effects in more detail gives some additional insights. Hjelmar (2010) for example differentiated availability and visibility as two different factors and found that visibility was especially relevant for occasional buyers of organic food that did not plan to buy organic when entering the supermarket. For them being confronted with a presentation of organic produce that cannot be overlooked made the difference. Presentation at eye level, right next to the conventional alternative was what this segment of the customers reacted positively to.

Documented effects of shelf placement on product choice in a supermarket can also be used to increase sales of organic products. The more shelf space an item receives the more likely is it that it is selected (Desment & Renaudin, 1998; Dreze et al., 1994). In the supermarket the shelf space is usually distributed in strong disfavour of organically produced products. A position on eye or hand level also has a strong positive effect on sales numbers (Campo & Gijsbrechts, 2005). Items presented earlier on the shelf as well as near focal items (items highly preferred) also tend to have higher likelihoods of being sold (Simonson and Winer, 1992). Such effects are especially relevant, when the customers are under time pressure or are not motivated to engage in the shopping decision (e.g., after work shopping). Interestingly the effect of shelf-placement has been under-researched when it comes to organic food. One of the very few exceptions is a very comprehensive study by van Nierop et al. (2010). They found the best market share of organic products when they were presented in the middle of the shelf space and at eye level. They furthermore found that placing all organic food products in one corner of the supermarket does not increase sales but sorting the whole product category by brand (organic as well as conventional) does.

3.3 Price

Many studies found that higher prices for organic produce are the main barrier named by customers when asked why they do not buy organic (see Hughner et al, 2007, for a review). The relation is, however, more complex than it appears at first glance. When asked, if they are willing to pay a premium for organically grown food, consumers usually state that they are (Batte et al., 2007). Interestingly, the amount people state they are willing to pay as a premium is in many food categories lower than the actual premium (Millock et al., 2002) and this might not be a coincidence: Stating to be willing to pay a premium but at the same time naming an amount for the acceptable premium that lies below the actual premium is a very convenient way to both keep a clear conscience ("I am willing to financially support organic farming...") and continue not buying ("... but the actual premium is too high"). Soler et al. (2002) however present an alternative explanation: Based on results from their experimental study they assume that the decision to pay a premium is two-fold, first a decision is made, if a premium should be paid or not. About 70% of their participants were willing to pay a premium. This decision is more determined by attitudes towards environment and food safety. Then a second decision is made on the amount of the premium that is acceptable. This decision is more determined by socio-economic variables.

Factors that have been shown to impact willingness to pay for organic products are a perceived added value with respect to food quality and security as well as trust in the producers and marketing chain (Krystallis & Chryssohoidis, 2005). Furthermore, willingness to pay increases with strong pro-environmental attitudes and young children in the household (Soler et al., 2002). They found that willingness to pay for an organic product is higher if a reference price for a conventional product is named and if information about the organic alternative is given orally (as opposed to written).

Interestingly, having to pay a premium on organic food is not only a barrier to purchase but also has a positive effect on the perception of the quality: Hill and Lynchehaun (2002) found that consumers used the price difference to infer that organic products both have better quality and taste. Also Cicia et al. (2002) demonstrated that customers used the price as a proxy to determine the quality of organic olive oil. Too low prices on organic olive oil were associated with it being of poor quality or not even truly organic.

The effect of the premium on organic food purchase is thus multi-dimensional. On the one hand it is a barrier that makes purchases less likely, on the other it is a proximal indicator associated with high quality food which might function as a motivator for a purchase - at least for some people. People willing to pay a price for higher quality food seem not to be scared off by the premium while people focussing on the budget are.

4. The importance of trust

Since the customers are not able to trace back their food through the whole production chain - at least not without considerable effort - trust in the farmers, producers and vendors becomes a key issue. This is especially the case for the organic food sector which is probably more than most other sectors depending on its costumers' trust. Brom (2000) analyses trust in the food sector and concludes that because of the de-coupling of food production and food consumption trust in food needs to be institutionalized. Trust, usually built in personal communication, needs to be established in another way. This means procedures of governmental (or other independent institutional) control in the food section need to be implemented to sustain consumer trust. Brom (2000) calls concerns about food safety (see section 2.3) an indicator of losing trust. He furthermore claims that trust is a moral relation, which means that there can only be trust in the food sector if the moral concerns of consumers are taken seriously.

There are literally hundreds of studies indicating that trust in producers and certifying institutions is one of the key determinants of organic food purchase (e.g., Krystallis & Chryssohoidis, 2005; Padel & Foster, 2005; Harper & Makatouni, 2002). Mistrusting that food marketed with organic food labels really is organic or that organic farming really makes a difference with respect to the food attributes important for a person (e.g., environmental friendliness, animal welfare, food safety, better taste, etc.; see section 2.2) is an almost certain death blow for any intention to buy organic food. Aarset et al. (2004) demonstrated that distrust in organic food certification is common in many countries and that this has a negative impact on attitudes towards organic food.

Following, a selection of factors will be described that have been shown to have a positive or negative impact on trust. Giannakas (2002) demonstrated in a mathematical model that keeping the amount of mislabelling conventional for organic produce low is vital for the organic food market. The organic food market will collapse if too many cases of mislabelling occur and customers' trust in organic labels is undermined. The importance of personal relations shows in a study by Sirieix and Schaer (2005). They found that French customers prefer to buy organic food on markets over supermarkets or health food stores because they experience a closer connection to their vendor, sometimes even communicating directly to the producer. Health food stores are trusted more than conventional supermarkets. In supermarkets trust is put in the food label, not the supermarket. Very similar results are presented by Essoussi and Zahaf (2009): They found in qualitative interviews that trust in organic food and food labels is the lower the longer the marketing chain is and the bigger the involved actors are. Direct marketing by local farmers receives the highest degree of trust, as consumers have direct access to information. Speciality stores receive a medium level of trust, because customer relations are perceived as being still rather close. Supermarkets receive the lowest trust rating and trust is transferred to the food label instead. Pivato et al. (2008) on the other hand found a relation between the perceived corporate social responsibilities of a supermarket chain impact the amount of trust customers have in their private label organic product series which eventually impacts also brand loyalty. Essoussi and Zahaf (2009) also found that organic food produced within your own country is trusted more than imported organic food. Truninger (2006) used in depth interviews to identify small size of the shop, personal bonding with the owner or personnel, feeling of belonging to one community with shared values and interests with the producer or vendor as determinants of trust. Furthermore, authenticity of organic food is validated by the appearance of the product: Fruits and vegetables should for example not be too big, too shiny, have small holes or bugs to be perceived as authentically organic. These findings put an interesting ambiguity on the appearance factor of organic food: on the one hand customers name sensory defects as barrier towards purchase of organic food (Hugher et al., 2007), on the other hand are exactly these sensory defects used as indicators of authenticity by other people. Maybe, the difference lies in the market segment: occasional buyers of organic fruits and vegetables in supermarkets expect the same visual appearance from organic than from conventional products, whereas more frequent buyers of organic food get suspicious when presentation is too perfect and shiny.

5. Models of environmental behaviour applied to organic food purchase

Social psychological behaviour models have contributed significantly to understanding environmentally relevant behaviour, its determinants and entry points for interventions to change behaviour. The two models most prominently used in environmental psychology today are the theory of planned behaviour (Ajzen, 1991) and the norm-activation theory (Schwartz & Howard, 1981). The two following sections outline their main assumptions and how the models have been used with respect to organic food purchase.

5.1 The theory of planned behaviour

The theory of planned behaviour (Ajzen, 1991) was not specifically developed to explain organic food choice but all types of planned behaviour. Its main assumption is that behaviour usually is under volitional control and is in such a case guided mainly by the intention to perform it: an actor develops a will to perform a certain behaviour (e.g., buying organic milk) and perceives it likely that this will happen. This intention is itself determined by three different factors (see figure 2): (a) the attitude towards the behaviour, (b) social norms, and (c) perceived behavioural control. The attitude is the sum of all beliefs about the behaviour (see section 2.2). Beliefs are expected outcomes times evaluation of the outcomes which makes the attitude a measure of favourability of a behavioural option. Social norms - they were called subjective norms by Ajzen (1991) - are the perceived expectations of other people: What do I think other people expect me to do in this situation? Is buying organic milk socially acceptable? Would people that are important to me support me in doing that? Would they expect it and probably sanction me for not doing it? Are other people's expectations important to me for this particular behaviour? Social norms have been further separated into injunctive and descriptive norms (Thøgersen, 2006). Injunctive norms are the anticipated expectations of other people about what is right and what is wrong, what is appropriate and what is not. Descriptive norms on the other hand are simply a representation of what other people do. Perceived behavioural control, finally, is the degree of control a person experiences over his or her behaviour. Is it easy for me to buy organic milk? Are there external factors that prevent me from doing it (e.g., availability, restricted budget)? Under certain conditions perceived behavioural control can affect behaviour directly and shortcut the mediation by intentions (see figure 2).

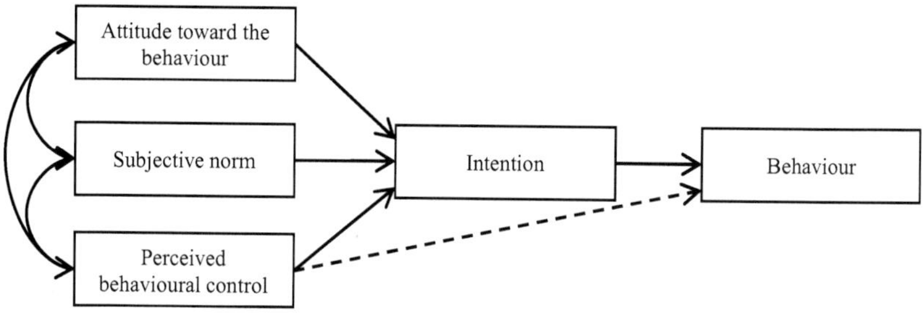

Fig. 2. The theory of planned behaviour (Ajzen, 1991, p. 182)

In the domain of organic food purchase the theory of planned behaviour has been applied successfully by several authors. Selected examples are presented in the remainder of this section, all extending the theory of planned behaviour by additional aspects. Arvola et al. (2008) used the theory as a framework to test factors that influence intentions to buy organic food (apples and pizza) in three European countries (Italy, Finland, and UK). They extended the model by "moral attitudes", which they define as "*the self-rewarding feeling of doing the right thing*" (page 443). For the intention to buy organic apples, social norms and attitudes were the only predictors, whereas moral attitudes became a third predictor of intentions for organic pizza. Interestingly, perceived behavioural control came out as not related to intentions in their study and was consequently omitted from the model. The strongest between countries differences were that the moral attitude was a stronger predictor than social norms in Italy and the UK, whereas in Finland it was the other way round. This indicates that the impact the factors in the theory of planned behaviour have on intention is depending on culture. Tarkiainen and Sundqvist (2005) on the other hand found evidence in a different Finnish sample that the impact of social norms is only indirect and mediated by attitudes. Additional factors like health consciousness, importance of price and perception of availability did not impact the self-reported purchase frequency significantly.

Vermeir and Verbeke (2008) tested a model where they assumed that attitudes and social norms impact intentions to buy a hypothetical organic diary product. They divided perceived behavioural control into two sub-dimensions, one being perceived availability and one perceived consumer effectiveness. The latter captures if people feel they - as consumers - can make a noticeable difference. Furthermore, they tested if value orientations (see section 2.1) and confidence that the product does what it promises, which could be interpreted as a measure of trust, moderate the relations between the four predictors and intentions. They found that attitudes had the by far strongest influence on intentions, followed by the two sub-dimensions of perceived behavioural control. Social norms only had a weak influence. Value orientations did moderate the relations between social norms and intentions as well as between perceived consumer effectiveness and intentions. Social norms have a stronger influence for people with low scores on the universalism and stimulation value and high scores on tradition and self-direction. Perceived consumer effectiveness is less important for people with high traditional values and low stimulation values. Social norms were not a significant predictor of intentions for people with low confidence that the product does what it promises.

5.2 The norm-activation theory

The norm-activation theory (Schwartz & Howard, 1981) in contrast to the theory of planned behaviour focuses on personal norms as the main driver of behaviour. It was developed to explain pro-social behaviour but has been adapted to environmentally relevant behaviour (e.g., Hunecke et al., 2001). The theory assumes that personal norms, which are a feeling of moral obligation to act in a certain way, predict behaviour directly. Obviously, this effect only applies to motivations that have a moral undertone (see section 2). People that buy organic food for hedonistic or health reasons would not be affected by moral obligations. To become relevant, personal norms have to be triggered in a situation when a decision is made. Activating factors in the model are the perception of ecological problems, awareness of consequences, subjective norms and perceived behavioural control (see figure 3).
Subjective norms and perceived behavioural control are identical to the theory of planned behaviour. Perception of problems captures that personal norms are only activated when a person perceives a problem to be relevant in a given situation. Awareness of consequences reflects the extent to which a person perceives his or her actions to contribute significantly to the problem. In their adaptation of the norm-activation theory to environmental behaviour, Hunecke et al. (2001) expected that the relation between personal norms and behaviour is moderated by external costs. Klöckner and Ohms (2009) applied the model to the purchase of organic milk and found support for the relations suggested in the model. Thøgersen and Ölander (2006) found in a panel study that strong personal norms are a good predictor of changes in consumption patterns towards organic products. The impact of perceived consumer effectiveness (see section 5.1) on behaviour is mediated by personal norms.

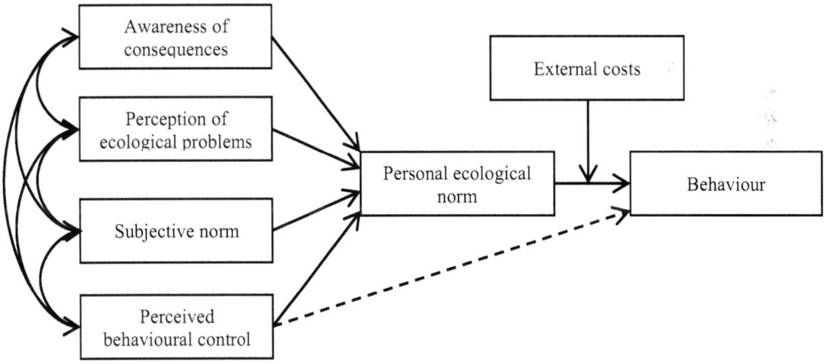

Fig. 3. The norm-activation theory (as adapted by Hunecke et al., 2001, p. 832)

6. Models of health behaviour applied to organic food purchase

As some consumers buy organic food out of health concern or concern for food safety, it makes sense to analyse briefly how health psychological models might contribute to understanding organic food choice. The health-belief model (e.g., Rosenstock, 2000) and the related protection-motivation theory (R. W. Rogers, 1975) have been applied to organic food purchase. The protection-motivation theory assumes that a motivation to protect oneself against health or other threats results from the evaluation of two factors: (a) how big is the threat for me personally and (b) how effective are coping measures I can take. The threat appraisal depends on the perceived severity of the threat and the perceived vulnerability. Highest threat appraisals occur therefore when a threat is connected to severe consequences and a person considers him/herself to be vulnerable. The threat appraisal can be reduced if the threat is connected to some kind of behaviour which is intrinsically or extrinsically rewarded (e.g., eating sweets). A high threat appraisal alone is however not enough to motivate protection measures. In addition a person has to come to the conclusion that coping strategies are effective in reducing the threat (response efficacy), feasible (self-efficacy) and not too costly. Taking organic meat consumption as an example, a motivation to buy organic meat would develop if a person perceives a relevant threat with severe consequences (e.g., being infected with Creutzfeld-Jakob disease when eating BSE infected meat), perceives herself as being vulnerable (e.g., being a frequent meat eater), perceiving the option to buy organic meat as effective (e.g., no infection with BSE in organic meat), feasible (e.g., there is organic meat sold in the local supermarket) and not too costly (e.g., premium for organic meat is affordable).

Verhoef (2005) used variables of the protection motivation theory to explain preference for organic meat and found that fear of health related consequences of consumption of conventional meat is a relevant predictor. Scarpa & Thiene (2011) used the protection motivation theory constructs to identify sub-groups of Italian people buying organic carrots. Based on protection motivation theory they identified three classes of people: (a) the first and with 60% largest class consisted of people that had both a high threat appraisal (threat of pesticide residues in conventional carrots) and a high coping appraisal (buying organic carrots helps and is feasible), (b) a second class of 25% with high coping appraisal but low threat appraisal (which should show some action, "just to be sure"), and (c) a small class of 15% with low threat and coping appraisal. A class with high threat appraisal and low coping appraisal was not found.

7. Perception and use of organic food labels

Food labels on organic food have been discussed in section 4 already as a trust-building aspect in the purchase of organic produce, especially if the purchase is made in an environment that is not trusted per se (e.g., a supermarket). All over the world, hundreds if not thousands of organic food labels exist, varying a lot in what they certificate and who the administering authority is. You can find labels only valid for certain lines of products (e.g., specific organic wine labels), labels that are used across the whole range of food products, labels that are assigned only in one country or region, labels that are used across country borders, labels that are assigned by independent organizations or governmental organizations, labels that are assigned by the food industry itself or interest organizations, additional organic food labels that are supermarket chain specific and so on. The standards for each label are different so that organic products often carry a selection of several labels (e.g., the general European Union organic food label in addition to the local label with stricter standards). This large variation leads to potential confusion of customers about standard behind eco-labels and mistrust might be a result. In a review article Pedersen & Neergard (2006) show that a large majority of consumers indicate that there were too many labelling schemes. Furthermore, most people indicated very limited knowledge about what the labels actually stand for and even about some of the basic concepts involved.

Teisl and Roe (2005) summarize the factors that contribute to effective eco-labelling programmes. First of all, customers have to notice, understand and belief the information communicated by the label. Since customers do not have the means to verify that a certain product actually fulfils the standards that the eco-label promises, the "belief" is a matter of trust and credibility of the certifying institution, which has been discussed in section 4. "Notice" addresses the problem that the eco-label is competing for customers' attention with many other labels, logos and visual influences in the supermarket.[1] To become relevant in a decision it has to be - at least subconsciously - noticed. Furthermore, the customer needs to be able to connect the label with a message relevant for him or her (e.g., this produce is from organic farming). Teisl and Roe (2005) found in a series of experiments with differently designed eco-labels that their perceived credibility was higher if contact information was added, if more detailed numbers instead of summary scores were presented, if the certificating organization was familiar to the customer, independent from the producer, and visible with a logo close to the label or in the label. Biel and Grankvist (2010) also found in a study with professional food purchasers that more detailed information positively impacted the choice of the more environmentally friendly product. Teisl and Roe (2005) were also able to show that credible labels had an effect on product choice. Tang et al. (2004) analysed the impact of visual and verbal communication on eco-labels and found that both had an independent and additive effect, meaning that combining visual and verbal communication had the largest effect. Søderskov and Daugbjerg (2011) were able to show that trust in eco-labels is higher in countries with where the state is more involved in assigning eco-labels.

Leire and Thidell (2005) outline in a review paper that Nordic customers are to a large extent very aware of eco-labels: they recognize them, know about their background and trust the certifying authorities sufficiently. However, a much smaller proportion of the Nordic population actually buys products with these labels. Leire and Thidell (2005) conclude that the use of eco-labels in the dynamics of and in interaction with the choice situation in the supermarket is under-researched.

Grankvist et al. (2004) experimentally compared the effect of positive labels and negative labels. Positive labels indicate the benefit the use of a product has for the environment compared to an average product; negative labels indicate the increased negative outcomes the use of a product would have compared to an average product from that category. They found an interesting interaction between the effect of positive and negative labels and the consumers environmental interest: people with low environmental interest were not affected by any type of label, people with high environmental interest reacted to both types, but individuals with an intermediate interest reacted more strongly to the negative label. The effect that negative information had stronger effects than positive was also replicated by Biel and Grankvist (2010). Given that almost all food labels are positive labels this finding indicates that negative labels on especially environmentally damaging products could reach a higher proportion of the population - not taking the feasibility of that approach in current market conditions into account.

---

[1] When the consumer decides to shop in an organic food speciality store organic food labels only have a reduced importance: to select between different organic food standards.

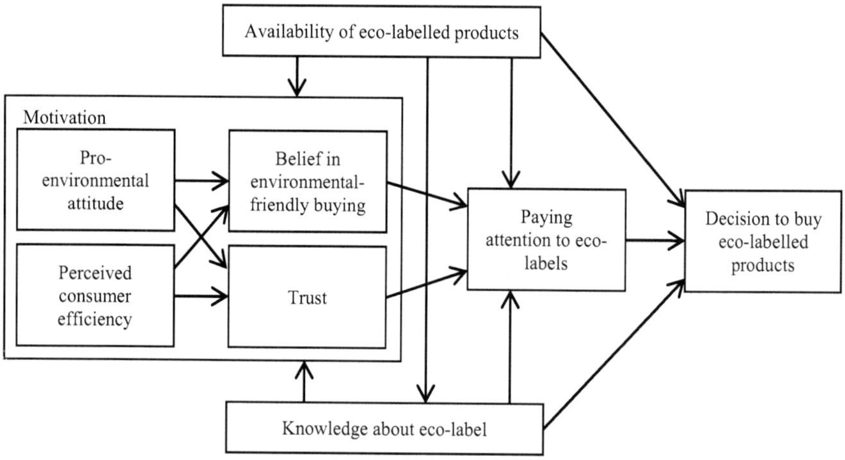

Fig. 4. A model of predicting paying attention to eco-labels and the purchase of labelled products (Thøgersen, 2000, p. 293)

Thøgersen (2000) developed a framework model that describes under which conditions consumers pay attention to eco-labels (see figure 4). He assumes that a decision to buy an eco-labelled product in the supermarket is depending on availability of labelled products in the store and knowledge about the label but also on paying attention to the labels.

Following the model paying attention is influenced by availability, knowledge, a fundamental belief in benefits of environment-friendly buying and trust. The latter two are impacted by a general pro-environmental attitude and perceived consumer effectiveness (see section 5.1). Thøgersen (2000) tested parts of the proposed models on a sample of customers from five European countries and found general support for the model.

In a recent paper Thøgersen et al. (2010) applied E. M. Rogers (1995) diffusion of innovation theory to the adoption of eco-labels. Based on the theory they developed a framework model of the adoption process of an eco-label and how it diffuses through a population (see figure 5). They assume that the individual process of adoption goes through six stages: (a) the individual needs to be exposed to the new label, (b) the individual needs to perceive it at least subconsciously, (c) the individual needs to understand the label and its message and needs to make inferences about what it means related to goals that are important for the individual, (d) the individual evaluates the message and potentially likes it, (e) the product is tried once, and if that resulted in satisfaction (f) adoption becomes more permanent.

The speed of this process depends on factors within the environment (e.g., how much effort is put into campaigning or how many other people already adopted the label), the adopting person, and the label itself. Using a food label for sustainable fish as an example they identified factors that contributed to start the adoption process (perceiving the label and aiming to understand it) and factors that contribute to complete the process (trying and continuing to use the label for purchase decisions). General knowledge about eco-labels, subjective knowledge about sustainable fishery, having the intention to buy sustainable fish, the degree of innovativeness with respect to eco-labels, and being female contributed positively to the probability of starting the adoption process. Innovativeness captures if a person perceives him/herself to be an early adopter of organic food related innovations. Successfully coming through the first stages of perceiving, understanding and liking the label are good predictors of final adoption. The buying intention still has a direct effect, but the interaction with having passed the first stages is also significant, showing that participants that both intent to buy sustainable fish and recognized and understood the label are more likely to buy the labelled fish than people that only have the intention.

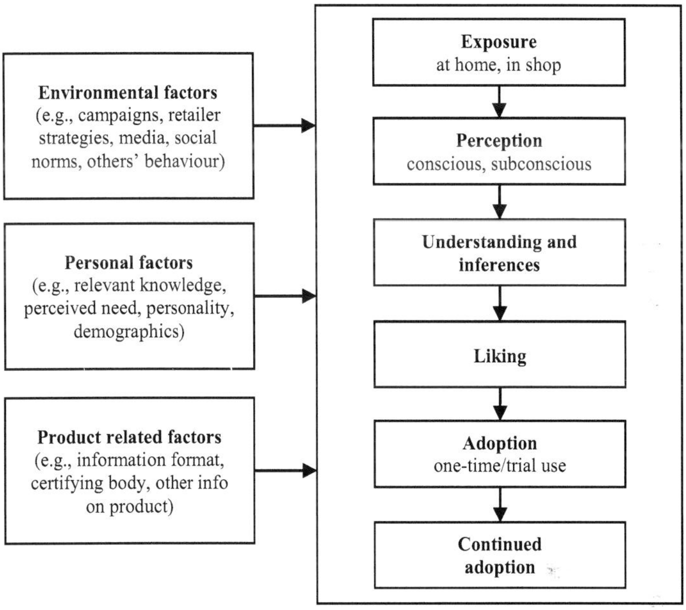

Fig. 5. A framework model of eco-label adoption (Thøgersen et al., 2010, p. 1790)

8. An integrated framework: In the supermarket and beyond

Building on the finding presented in the previous sections and a comprehensive model of environmental behaviour proposed by Klöckner (2010) an integrated modelling framework is suggested. Despite its complexity it is not meant to be a complete model of consumer behaviour but a framework to analyse the subtleties and interplay of the variables that have been introduced before. The first assumption of the model is that consumer behaviour with respect to organic food is not the result of one decision but a series of decisions nested in each other. As an example two of those decisions and possible determinants are depicted in figure 6: (a) the decision where to go for food shopping (for reasons of keeping the model reasonably simple only with the alternatives speciality organic food store and supermarket) is displayed in the upper half, (b) if the first decision is for the supermarket, more decisions have to be made between conventional and organic products within the supermarket. If a decision is made for a speciality store, the following in-store decisions do not affect the outcome with respect to the broad categories organic vs. conventional food. Of course decisions are made also in speciality stores and of course also within the category of organic food these decisions shape the environmental impact, but this is deliberately left out of further analysis.

Let us take a closer look at the shopping location decision and its determinants first. According to the action models presented in section 5 and the arguments presented by Klöckner (2010) this decision should be impacted by three different variables: Intentions to buy in a speciality store, perceived control over this behaviour and shopping habits or store loyalties. Shopping habits refers to if people repeatedly did their shopping in a particular store so that the decision where to go for a shopping trip might be shortcut and people go just where they usually go. This effect might be in favour or disfavour of the organic food store. Habits should reduce the impact of intentions on the choice, especially if people decide under time pressure or with low emotional involvement. Perceived behavioural control is divided into three sub-dimensions: (a) perceived consumer effectiveness, (b) perceived availability of a speciality store, and (c) perceived convenience of shopping there (e.g., how do I get there, do they offer everything I need, is it on my way to other activities, etc.). Also perceived behavioural control should not only impact the decision but also the strength of the impact of intentions: If perceived behavioural control is low, the impact of intentions on behaviour should be reduced. Finally, mistrust in the credibility of the food store might interfere with the intention to buy there.

Intentions to buy in a speciality store should be affected by the attitudes towards organic food and speciality stores, personal norms (which also might be called moral attitudes), social norms, and for some people also protection motivations out of health concern. Attitudes are built on beliefs, personal norms are a reference to value orientations, and social norms can be divided into injunctive norms (what people say to other people what they should do) and descriptive norms (what other people do). Finally, protection motivation is determined by the appraisal of a possible threat connected to shopping in a conventional supermarket and the coping appraisal.

Many variables and relations in the lower part (the in-store decisions) are similar to the variables in the upper half, but it is important to keep in mind, that they refer now to a different decision: Intentions are now intentions to buy the organic version of one specific product, the attitudes are attitudes about this specific product, habits are now routines in the shop (for example which way to go through the aisles, in which and where on the shelf to look for products, which products to prefer automatically) and brand loyalties. Perceived control is also specific for this decision and incorporates specific versions of perceived consumer efficiency, the availability and visibility of a product and the premium that has to be paid. Social norms and a potential protection motivation are also connected specifically to products or product classes. All of these variables will differ from the more general ones described before and also between product categories and products.

Another important difference between the in-store decision and the between stores decision is that eco-labels become a central position in enabling people to act according to their intention to buy organic food. Organic food has to be identified and usually food labels make that possible. As has been described before, people pay attention to food labels if they intent to buy organic, but also only if labelled products are available and visible in the shop, if the label is trusted and familiar. Visibility is affected by marketing within and outside the store but also where and how a product is presented on the shelves. Visibility affects familiarity, which in turn also affects trust (the more familiar the more trusted).

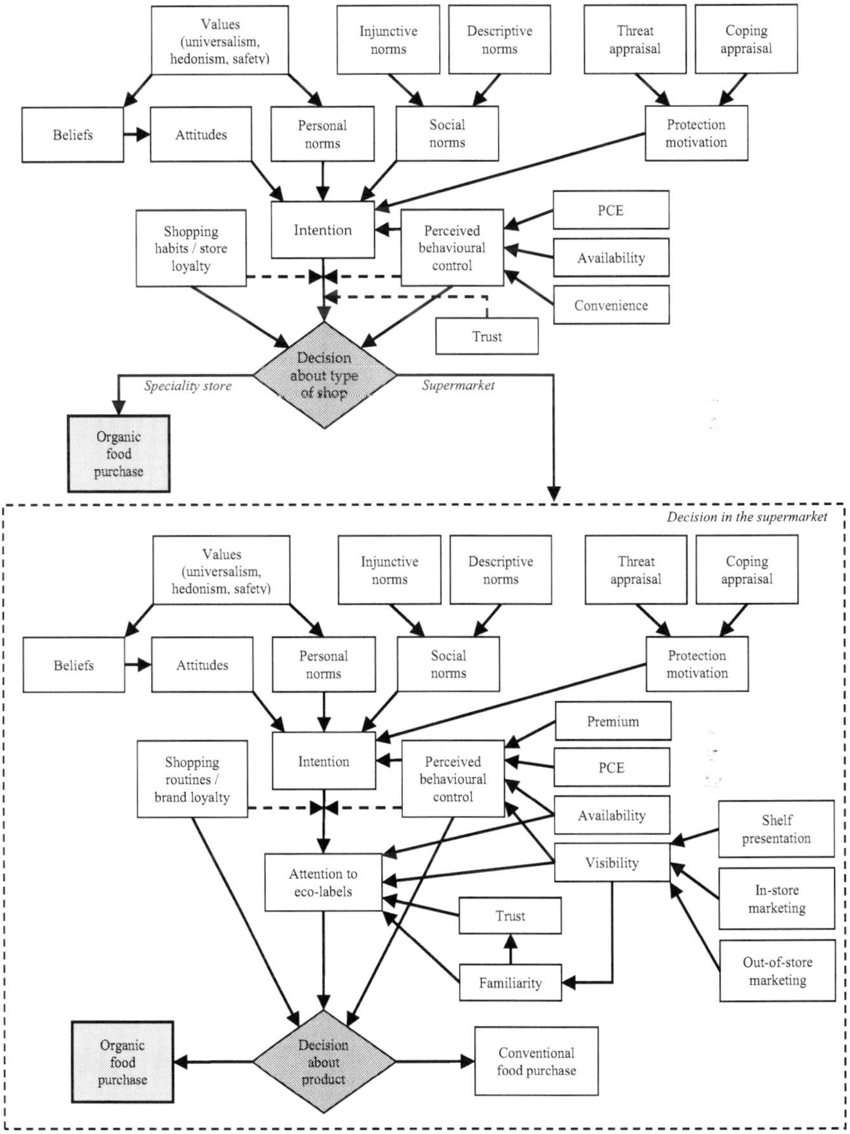

Fig. 6. An integrated model framework

9. Conclusion

In this chapter the complexity of human decision-making with respect to purchasing organic food has been outlined. It has been demonstrated that enhancing the diffusion of organic food in the market is more complex than just lowering the absolute price for organic food. Even if high prices for organically produced food have been repeatedly identified as a major barrier in the purchase process (e.g., Chinnici et al., 2002), the price of a product is only one determinant of a purchase decision. To understand the role the price can or cannot play in the decision it has to be differentiated between the absolute and the relative price. If only the absolute price is evaluated it is difficult for a consumer to decide, if a product is too expensive to be purchased or not. The absolute price can only be evaluated against the available budget which poses an upper limit to the expenses that can be made or a very abstract scale of what appears to be a high price within the category of food. Therefore, customers usually determine more accurately if a product is expensive or not based on relating its price to a reference price, in case of organic food to a similar conventionally produced product (Soler et al., 2002). This relative price should then have much more relevance for purchase decisions than absolute prices as long as the absolute price does not overstretch the budget and the product is outside the range of the affordable. This argument can be underlined by looking at studies analysing the impact of increases or decreases of the absolute price level of a product category compared to changes in the relative pricing structure within a category. Whereas for many products at least moderate increases of the absolute price level often have no effect on the quantity of product purchases - an effect referred to as price inelasticity - changes of the relative price structure does, for example during promotional campaigns (Bolton, 1989). However, even if we accept that within the boundaries of the available budget the relative price may be more important than the absolute the presented framework model suggests that the price is only one of many determinants of purchasing organic food or not. Availability and visibility are often at least as important, especially in societies with a high average income level that spend a rather low proportion on food purchases and especially for non-committed buyers. Moral or health protective motivations are relevant, though not directly impacting purchase behaviour. A motivation to buy organic is fragile and can easily be forgotten or deactivated by other motivations on the way into the supermarket.

The framework model presented in the previous section offers various potential levers to impact the market share of organic products. The various motivations to consider organic food have been presented, possible barriers have been identified and the aspect of communicating with the customer via labels has been analysed. Given that the most potential growth sections for organic food lie in the supermarket and not the speciality store (Sahota, 2007), some recommendations based on the model will be presented in this section: to be purchased by a customer in a supermarket organic food has to be available and visible when the decision is made. Shelf placement of and space occupied by organic food plays a crucial role. As long as the price premium is not too extensive the premium is no insurmountable barrier. On the contrary, it also carries the message of high quality food. With respect to food labels visibility, tangibility and trust to the administering authority are the important aspects. Only if all of these things are in place in the supermarket motivations to protect nature, animals, one's health or producers have the chance to become translated into behaviour. An important additional condition is, that the consumer perceives the contribution made by this particular purchase relevant for solving the moral or health related dilemma. Thus, communication the positive impact a purchase can have (either individually or aggregated to a meaningful number) is a promising strategy. Finally, change in purchasing habits is a difficult process often interfered by brand loyalties or automaticities in the supermarket. Breaking such habits often affords structural changes like rearranging the supermarket layout, a technique often applied by supermarket chains to prevent customers from developing too powerful shopping routines that would reduce their perceptiveness for new products.

Although the framework model is probably too complex for a model test, it makes interesting predictions that can be tested. Especially the nested structure of decision and the impact earlier decisions have on the decisional space of later decisions is very much under-researched and should get more attention. Also the social context of food purchases which means the direct or indirect information customers detect and process about what other people buy and what is normal and accepted should get more attention. Finally, the application of conventional marketing strategies on organic food and studies about if and how they apply to this special food sector is still pending. Very little is known about if traditional rules for shelf placement or shelf space occupied function in the same way for organic food than they do for conventional.

10. References

Aarset, B., Beckmann, S., Bigne, E., Beveridge, M., Bjorndal, T., Bunting, J., McDonagh, P., Mariojouls, C., Muir, J., Prothero, A., Reisch, L., Smith, A., Tveteras, R., & Young, J. (2004). The European consumers' understanding and perception of the "organic" food regime - the case of aquaculture. *British Food Journal*, 106, 2, ISSN 0007-070X, 93-105

Ajzen, I. (1991). The theory of planned behavior. *Organizational Behavior and Human Decision Processes*, 50, 2, ISSN 0749-5978, 179-211

Armitage, C. J., & Christian, J. (2003). From attitudes to behaviour: Basic and applied research on the theory of planned behaviour. *Current Psychology*, 22, 3, ISSN 1046-1310, 187-195

Arvola, A., Vassallo, M., Dean, M., Lampila, P., Saba, A., Lähteenmäki, L., & Shepherd, R. (2007). Predicting intentions to purchase organic food: The role of affective and moral attitudes in the Theory of Planned Behaviour. *Appetite*, 50, 2-3, ISSN 0195-6663, 443-454

Aschemann, J., Hamm, U., Naspetti, S., & Zanoli, R. (2007). The organic market, In: *Organic Farming : an International History*, W. Lockeretz, 123-151, CABI, ISBN 978-0-85199-833-6, Wallingford, England

Batte, M. T., Hooker, N. H., Haab, T. C., & Beaverson, J. (2007). Putting your money where their mouths are: Consumer willingness to pay for multi-ingredient, processed organic food products. *Food Policy*, 32, 2, ISSN 0306-9192, 145-159

Baker, S., Thompson, K. E., Engelken, J., & Huntley, K. (2004). Mapping the values driving organic food choice: Germany vs. The UK. *European Journal of Marketing*, 38, 8, ISSN 0309-0566, 995-1012

Biel, A., & Grankvist, G. (2010). The effect of environmental information on professional purchasers' preference for food products. *British Food Journal*, 112, 3, ISSN 0007-070X, 251-260

Bolton, R. N. (1989). The relationship between market characteristics and promotional price elasticities. *Marketing Science*, 8, 2, ISSN 1526-548X, 153-169

Brom, F. W. A. (2000). Food, consumer concerns, and trust: food ethics for a globalizing market. *Journal of Agricultural and Environmental Ethics*, 12, 2, ISSN 1573-322X, 127-139

Campo, K., & Gijsbrechts, E. (2005). Retail assortment, shelf and stockout management: Issues, interplay and future challenges. *Applied Stochastic Models in Business and Industry*, 21, 3, ISSN 1526-4025, 383-392

Chinnici, G., D'Amico, M., Pecorino, B. (2002). A multivariate statistical analysis on the consumers of organic products. *British Food Journal*, 104, 3-5, ISSN 0007-070X, 187-199

Cicia, G., del Giudice, T., & Scarpa, R. (2002). Consumers' perception of quality in organic food - a random utility model under preference heterogeneity and choice correlation from rank-orderings. *British Food Journal*, 104, 3-5, ISSN 0007-070X, 200-213

De Magistris, T., & Gracia, A. (2008). The decision to buy organic food products in Southern Italy. *British Food Journal*, 110, 9, ISSN 0007-070X, 929-947

Desmet, P., & Renaudin, V. (1998). Estimation of product category sales responsiveness to allocated space. *International Journal of Research in Marketing*, 15, 5, ISSN 0167-8116, 443-457

Dreezens, E., Martijn, C., Tenbült, P., Kok, G., & de Vries, N. K. (2005). Food and values : an examination of values underlying attitudes toward genetically modified- and organically grown food products. *Appetite*, 44, 1, ISSN 0195-6663, 115-122

Dreze, X., Hoch, S. J., & Purk, M. E. (1994). Shelf-management and space elasticity. *Journal of Retailing*, 70, 4, ISSN 0022-4359, 301-326

Eagly, A. H., & Chaiken, S. (1993). *The Psychology of Attitudes*, Harcourt, Brace, Jovanovich, ISBN 0155000977, Fort Worth, Texas, USA

Ergin, E. A., & Ozsacmaci, B. (2011). Turkish consumers' perceptions and consumption of organic foods. *African Journal of Business Management*, 5, 3, ISSN 1990-3839, 910-914

Essoussi, L. H., & Zahaf, M. (2009). Exploring the decision-making process of Canadian organic food consumers - motivations and trust issues. *Qualitative Market Research : An International Journal*, 12, 4, ISSN 1352-2752, 443-459

Fishbein, M., & Ajzen, I. (1975). *Belief, Attitude, Intention, and Behavior: An Introduction to Theory and Research*, Addison-Wesley, ISBN 0201020890, Reading, Massachusetts, USA

Fox, J. (2006). Notice how you feel: An alternative to detached concern among hospice volunteers. *Qualitative Health Research*, 16, 7, ISSN 1552-7557, 944-961

Giannakas, K. (2002). Information asymmetries and consumption decisions in organic food product markets. *Canadian Journal of Agricultureal Economics*, 50, 1, ISSN 1744-7976, 35-50

Grankvist, G., Dahlstrand, U., & Biel, A. (2004). The impact of environmental labelling on consumer preference: Negative vs. positive labels. *Journal of Consumer Policy*, 27, 2, ISSN 0168-7034, 213-230

Grunert, S. C., & Juhl, H. J. (1995). Values, environmental attitudes, and buying of organic food. *Journal of Economic Psychology*, 16, 1, ISSN 0167-4870, 39-62

Harper, G. C., & Makatouni, A. (2002). Consumer perception of organic food production and farm animal welfare. *British Food Journal*, 104, 3-5, ISSN 0007-070X, 287-299

Hill, H., & Lynchehaun, F. (2002). Organic milk: Attitudes and consumption patterns. *British Food Journal*, 104, 7, ISSN 0007-070X, 526-542

Hjelmar, U. (2010). Consumers' purchase of organic food products. A matter of convenience and reflexive practices. *Appetite*, 56, ISSN 0195-6663, 336-344

Honkanen, P., Verplanken, B., & Olsen, S. O. (2006). Ethical values and motives driving organic food choice. *Journal of Consumer Behaviour*, 5, 5, ISSN 1479-1838, 420-430

Hunecke, M., Blöbaum, A., Matthies, E., & Höger, R. (2001). Responsibility and environment - ecological norm orientation and external factors in the domain of travel mode choice. *Environment and Behavior*, 33, 6, ISSN 0013-9165, 830-852

Klöckner, C. A. (2010). Understanding the purchase of organic food - a theoretical framework from a psychological perspective. *FoodInfo Online Features*, 23 August 2010: www.foodsciencecentral.com/fsc/ixid15967

Klöckner, C. A. (2008). *Subjektive barrierer som forhindrer at nordmenn handler økologisk mat.* Unpublished pilot study report, NTNU, Trondheim, Norway.

Klöckner, C. A., & Ohms, S. (2009). The importance of personal norms for purchasing organic milk. *British Food Journal*, 111, 11, ISSN 0007-070X, 1173-1187

Krystallis, A., & Chryssohoidis, G. (2005). Consumers willingness to pay for organic food - factors that affect it and variation per organic product type. *British Food Journal*, 107, 5, ISSN 0007-070X, 320-343

Lea, E., & Worsley, T. (2005). Australians' organic food beliefs, demographics and values. *British Food Journal*, 107, 10-11, ISSN 0007-070X, 855-869

Leire, C., & Thidell, Å. (2005). Product-related environmental information to guide consumer purchases - a review and analysis of research on perceptions, understanding and use among Nordic consumers. *Journal of Cleaner Production*, 13, 10-11, ISSN 0959-6526, 1061-1070

Magnusson, M. K., Arvola, A., Hursti, U., Åberg, L., & Sjöden, P. (2003). Choice of organic food is related to perceived consequences for human health and to environmentally friendly behaviour. *Appetite*, 40, 2, ISSN 0195-6663, 109-117

Makatouni, A. (2002). What motivates consumers to buy organic food in the UK? Results from a qualitative study. *British Food Journal*, 104, 3-4-5, ISSN 0007-070X, 345-352

Michaelidou, N., & Hassan, L. M. (2008). The role of health consciousness, food safety concern and ethical identity in attitudes and intentions towards organic food. *International Journal of Consumer Studies*, 32, 2, ISSN 1470-6431, 163-170

Millock, K., Gårn Hansen, L., Wier, M., & Mørch Andersen, L. (2002). *Willingness to pay for organic foods: A comparison between survey data and panel data from Denmark.* Research report retrieved online: http://orgprints.org./1754/

Nierop, E. van, Herpen, E. van, & Sloot, L. (2010). *The relationship between in-store marketing and observed sales of sustainable products: A shopper marketing view.* Research report, University of Groningen, Netherlands

Nygård, B., & Storstad, O. (1998). De-globalization of food markets ? Consumer perceptions of safe food: the case of Norway. *Sociologia Ruralis*, 38, 1, ISSN 0038-0199, 35-53

Onyango, B., Hallman, W. K., & Bellows, A. C. (2007). Purchasing organic food in US food systems : A study of attitudes and practice. *British Food Journal*, 109, 5, ISSN 0007-070X, 399-411

Ott, S. L. (1990). Supermarket shoppers' pesticide concerns and willingness to purchase certified pesticide residue-free fresh produce. *Agribusiness*, 6, 6, ISSN 1520-6297, 593-602

Özcelik, A. Ö., & Ucar, A. (2008). Turkish academic staff's perception of organic foods. *British Food Journal*, 110, 9, ISSN 0007-070X, 948-960

Padel, S., & Foster, C. (2005). Exploring the gap between attitudes and behaviour - understanding why consumers buy or do not buy organic food. *British Food Journal*, 107, 8, ISSN 0007-070X, 606-625

Pedersen, E. R., & Neergaard, P. (2006). Caveat emptor - let the buyer beware! Environmental labelling and the limitations of 'green' consumerism. *Business Strategy and the Environment*, 15, 1, ISSN 0964-4733, 15-29

Pivato, S., Misani, N., & Tencatti, A. (2008). The impact of corporate social responsibility on consumer trust: the case of organic food. *Business Ethics: A European Review*, 17, 1, ISSN 0962-8770, 3-12

Rogers, E. M. (1995). *Diffusion of Innovations* (5th edition), Free Press, ISBN 978-0743222099, New York

Rogers, R. W. (1975). A protection motivation theory of fear appeals and attitude change. *Journal of Psychology*, 91, 1, ISSN 0022-3980, 93-114.

Rosenstock, I. M. (2000). Health belief model. In: *Encyclopedia of Psychology*, Vol. 4, A. E. Kazdin, pp. 78-80, Oxford University Press, ISBN 1-55798-653-3, Washington, USA

Sahota, A. (2007). Overview of the global market for organic food and drink, In *The World of Organic Aggriculture - Statistics and Emerging Trends 2007*, H. Willer & M. Yussefi (Eds.), pp. 52-55, International Federation of Organic Aggriculture Movements, ISBN 3-934055-82-6, Bonn, Germany

Schifferstein, H. N. J., & Oude Ophuis, P. A. M. (1998). Health-related determinants of organic food consumption in The Netherlands. *Food Quality and Preference*, 9, 3, ISSN 0950-3293, 119-133

Schofield, N. G., Green, C., & Creed, F. (2008). Communication skills of health-care professionals working in oncology - can they be improved. *European Journal of Oncology Nursing*, 12, 1, ISSN 1462-3889, 4-13

Schwartz, S. H. (1992). Universals in the content and structure of values: Theoretical advances and empirical tests in 20 countries. *Advances in Experimental Social Psychology*, 25, ISBN 0-12-015225-8, 1-65

Schwartz, S. H. (1994). Are there universial aspects in the structure and contents of human values? *Journal of Social Issues*, 50, 4, 19-45, ISSN 1540-4560

Schwartz, S. H., & Howard, J. A. (1981). A normative decision-making model of altruism, In *Altruism and Helping Behavior*, J. P. Rushton & R. M. Sorrentino (Eds.), pp. 89-211, Erlbaum, ISBN 978-0124630505, Hillsdale

Simonson, I., & Winer, R. S. (1992). The influence of purchase quantity and display format on consumer preference for variety. *Journal of Consumer Research*, 19, 1, ISSN 0093-5301, 133-138

Sirieix, L., & Schaer, B. (2005). *Buying organic food in France: shopping habits and trust.* Working paper 1/2005, Moisa, Montpellier, France, retrieved online: http://129.3.20.41/eps/othr/papers/0512/0512010.pdf

Søderskov, K. M., Daugbjerg, C. (2011). The state and consumer confidence in eco-labelling: organic labelling in Denmark, Sweden, the United Kingdom and the United States. *Agriculture and Human Values,* online first, ISSN 1572-8366, 1-11

Soler, F., Gil, J. M., & Sanchez, M. (2002). Consumers' acceptability of organic food in Spain - results from an experimental auction market. *British Food Journal,* 104, 8, ISSN 0007-070X, 670-687

Storstad, O., & Bjørkhaug, H. (2003). Foundations of production and consumption of organic food in Norway : Common attitudes among farmers and consumers? *Agriculture and Human Values,* 20, 2, ISSN 0889-048X, 151-163

Tang, E., Fryxell, G. E., & Chow, C. S. F. (2004). Visual and verbal communication in the design of eco-label for green consumer products. *Journal of International Consumer Marketing,* 16, 4, ISSN 0896-1530, 85-105

Tarkiainen, A., & Sundqvist, S. (2005). Subjective norms, attitudes and intentions of Finnish consumers in buying organic food. *British Food Journal,* 107, 11, ISSN 0007-070X, 808-822

Teisl, M. F., & Roe, B. (2005). Evaluating the factors that impact the effectiveness of eco-labelling programmes, In: *Environment, Information and Consumer Behaviour,* S. Krarup & C. S. Russell, pp. 65-90, Edward Elgar, ISBN 1-84542-011-X, Northampton, Massachusetts, USA

Truninger, M. (2006). *Exploring trust in organic food consumption.* Paper presented at the Joint Organic Congress, Odense, Denmark, 30-31 May 2006, available online: http://orgprints.org/7512/

Thøgersen, J. (2000). Psychological determinants of paying attention to eco-labels in purchase decisions: Model development and multinational validation. *Journal of Consumer Policy,* 23, 3, ISSN 0168-7034, 285-313

Thøgersen, J. (2006). Norms for environmentally responsible behaviour: an extended taxonomy. *Journal of Environmental Psychology,* 26, 4, ISSN 0272-4944, 247-261

Thøgersen, J. (2010). Country differences in sustainable consumption: the case of organic food. *Journal of Macromarketing,* 30, 2, ISSN 0276-1467, 171-185

Thøgersen, J., Haugaard, P., & Olesen, A. (2010). Consumer responses to eco-labels. *European Journal of Marketing,* 44, 11-12, ISSN 0309-0566, 1787-1810

Thøgersen, J., & Ölander, F. (2006). The dynamic interaction of personal norms and environment-friendly buying behaviour: a panel study. *Journal of Applied Social Psychology,* 36, 7, ISSN 0021-9029, 1758-1780

Verdurme, A., Gellynck, X., & Viaene, J. (2002). Are organic food consumers opposed to GM food consumers? *British Food Journal,* 104, 8, ISSN 0007-070X, 610-623

Verhoef, P. C. (2005). Explaining purchase of organic meat by Dutch consumers. *European Review of Agricultural Economics,* 32, 2, ISSN 0165-1587, 245-267

Vermeir, I., & Verbeke, W. (2008). Sustainable food consumption among young adults in Belgium: Theory of planned behaviour and the role of confidence and values. *Ecological Economics,* 64, 3, ISSN 0921-8009, 542-553

Zanoli, R., & Naspetti, S. (2002). Consumer motivations in the purchase of organic food : A means-end approach. *British Food Journal*, 104, 8, ISSN 0007-070X, 643-653

4

The Organic Food Market: Opportunities and Challenges

Leila Hamzaoui-Essoussi and Mehdi Zahaf
Telfer School of Management, University of Ottawa
Canada

1. Introduction

Nowadays, most environmental challenges that humanity is facing relate to unsustainable consumption patterns and lifestyles. Sustainability is seen in this context as a consumption pattern that meets the needs of present generations without compromising the needs of future generations (Bruntland, 1987). This is also related to basic needs such as food. The present food chain is mainly based on food scarcity, GMOs, use of pesticides and antibiotics, and industrialization of the agricultural system. Growing consumer demand for organic food (OF) is based on most of these facts (Davies et al., 1995; Chryssohoidis and Krystallis, 2005). Organic production combines best environmental practices, preservation of natural resources, animal welfare standards while ensuring no use of genetic engineering, pesticides, additives, or fertilizers; each stage of the organic food production being controlled and certified. On the other hand, there are some unique challenges to the cost and logistics of moving locally or regionally produced organic foods to the market. Of particular interest is the concept of food mileage[1] and the situation of small and medium size farms. At this time production of such farms is rather limited amounting to a few hundred tons. Such a volume will be of little interest to mainstream grocery chains. Moreover, consumers seem to be ambivalent about channels of distribution. Trust/mistrust emerge as an important factor in deciding not only where to buy OF products but even whether to buy OF products or not. Therefore, food mileage, price, and the certification process could contribute significantly to OF consumers' consumption decisions of OF products. Finally, the challenge that the organic food sector is currently facing is a gap in the knowledge that spans between the marketing system in place, the value chain, and the value delivery network in the organic food system.

This chapter introduces the current literature and current market realities of the OF industry and presents a supply-demand model. This model integrates both demand and supply side key factors and is built to answer the questions of what, how, where and why consumers buy organic. The authors also attempt to show how the combination of (1) behavioral factors such as knowledge and trust orientations, (2) lifestyle factors such as principle oriented standard of living and sustainability, and (3) local food/food mileage factors such as

---

[1] The distance food travels from the production site to the final consumer. The more food miles that attach to a given food, the less sustainable and the less environmentally desirable that food is.

support for the local economy and food's country of origin, interact and explain the complex organic food consumer behavior. Last, the chapter focuses on explaining the decision making process of organic food consumers by characterizing the differences between market clusters.

2. General trends

2.1 The organic food market

The organic market is moving from a niche market to a mainstream market within the agricultural industry, and was originated in the nineties (Agrifood Canada, 2011). It following a number of food scares in the conventional sector. The global market for organic products approximated US $18 billion in 2000 then US $23 billion in 2002 and has increased by 43% reaching US $33 billion in 2005, and US $50 billion in 2008 (Willer and Yussefi, 2007; Van Elzakker and Eyhorn, 2010). Double-digit growth rates were observed each year, except in 2009 because of the world economic crisis reducing investments and consumer buying power (Willer and Kilcher, 2011). Further, there are 633,891 farms managing 31 million hectares of "organic" land (Willer & Yussefi, 2007). More specifically, Oceania and Europe account for almost two-third of the world's organic land; 39% for the former and 23% for the latter. At the country level, Australia (11.8 million hectares), Argentina (3.1 million hectares), China (2.3 million hectares) and the US (1.6 million hectares) have the greatest organic areas. These figures[2] translate into a total of 130 countries producing certified organic food, 90 of which are developing countries presenting ideal environmental conditions for the development of satisfactory organic produce. There were almost 1.9 million organic producers in 2009, an increase of 31% since 2008, mainly due to a large increase in the production in India. As a matter of fact, 40% of the world's organic producers are in Asia, followed by Africa (28%), and Latin America (16%). In North America, Canada allocates 0.7 million hectares to organic production while the United States has 2 million hectares. This represents 7% of the world's organic agricultural land.

Although organic agriculture is now going mainstream, demand remains concentrated in Europe and North America. However, these two regions are not self-sufficient because production is not meeting demand. It is also obvious that the supply is not located where the demand is. Most of the demand is coming from Europe and North America. Hence, large volumes of organic imports, coming in from other regions, are used to balance the undersupply. In Europe, sales of organic products approximated € 18,400 million in 2009 (Willer and Kilcher, 2011). The largest market for organic products in 2009 was Germany (5.8 billion euros) followed by France (3 billion euros) and the UK (2 billion euros). US sales of organic products grew in 2009 by 5.3%, to reach 26.6 billion US dollars, representing 3.7% of the food market. On the Canadian front, the report of Agri-Food Canada in 2010, based on the 2008 sales of organic foods, concludes that the total Canadian organic market approximates CA $2 billion annually (Willer and Kilcher, 2011). Further to this, sales growth rates by Canadian provinces are distributed as follow: Alberta (44%), British Columbia (34%), Maritimes (34%), Ontario (24%), and Quebec (21%) (Macey, 2007).

[2] Survey conducted in 2009 by Research Institute of Organic Agriculture (FiBL) and the International Federation of Organic Agriculture Movements (IFOAM).

2.2 Sustainability, local food, and organics

Nowadays, sustainability is becoming one of the main social issues in the business field. Pressure from investors, cuttings on production costs (eg. Walmart), development of a positive image, and being able to charge more for organic foods is prompting large grocery chains to go sustainable (Saha and Darnton, 2005). The issue of sustainability is also analyzed by Jones et al. (2001), concluding that corporate social responsibility in the food retailing industry is translated in terms of support for local food producers, fair trade, healthy eating, commitment to organic products, and help for the local community. For many consumers, the support of local farmers is considered a socially responsible behaviour and partially reflects the belief that OF is locally grown (Hughner et al., 2007). An increasing number of organic shoppers emphasize that local foods and sustainability are in direct relation with their motivation to buy organic food (Zepeda and Deal, 2009). Environmentally conscious consumers are willing to pay a much higher price for sustainable products such as organic and locally-produced foods as ethical considerations are becoming important factors in their decision making process. This encourages organizations to embody corporate social responsibility. It is important to state that there are two sets of consumers: hardcore OF consumers and regular OF consumers (Hamzaoui and Zahaf, 2009). The latter type of consumer is seen as a consumer that buys OF for health or taste reasons, while the former is depicted as an active consumer buying OF for environmental and ethical reasons, along with some health reasons. In other words, the hard-core consumer is commited to the environment whereas the regular consumer is commited to personal health.

It is important to note that despite the fact that organics have gone mainstream, there is a new trend amongst hard-core consumers regarding the rapid growth of "industrial organics". This trend is based on a viral and emergent discontent among consumers regarding how the organic food system is evolving (Bean and Sharp, 2011). The main critics are not related to the key elements in the current definition of organics. On the contrary, these concerns are directly related to some economic, environmental and social ideals such as production systems, size of the operations, distribution systems and channels, and capital intensity. The by-product of this situation is what Bean and Sharp (2011) call alternative food systems (AFS). They examine two pathways for achieving sustainability, and propose a comparison among different types of local and organic food consumers in terms of attitudes about food, agriculture and the environment. This helps to understand alternative food consumer's preferences and how these preferences create new demand in the market. Hence, AFS are seen as sustainable and economically, socially, and environmentally more viable than standard systems. Innovative food systems, such as local farmers' markets, are based on low-carbon food distribution systems and could be also classified as AFS. The slow food movement is another good example of AFS.

3. Supply side factors

3.1 Channels of distribution

The organic food industry is steadily moving from niche markets, e.g., small specialty shops, to mainstream markets, e.g., large supermarket chains (Jones et al., 2001; Tutunjian, 2008). Ten years ago the bulk of OF sales were made in specialty stores (95%) while the remaining 5% were realized in mainstream stores. Nowadays, the trend has been reversed (Organic Monitor, 2006). In some countries, distributors are promoting their own line of OF products under specific brand names (Rostoks, 2002; Tutunjian, 2004). Alternative distribution channels are being used and are characterized by a direct link between the producer and the consumer, eg. farmers' markets (Smithers et al., 2008).

In the United States organic meat and dairy are experiencing the highest growth rates, 55.4% and 23.5% respectively (Willer and Kilcher, 2011), while organic flowers and pet food saw the highest growth rates for non-food categories. Conversely, in Canada and according to Macey (2007), total mass market sales of certified OF products approximated CA $586 million allocated as follow CA $175 through small grocery stores, drug stores, and specialty stores, and CA $411 in large grocery chains. These figures do not account for the alternative distribution channels such as farmers' markets, natural food stores, box delivery, and other channels such as restaurants. These channels totalize CA $415 million (Macey, 2007). This is also related to the structure of the current distribution systems. Hence, the pattern described in the previous section is clear. There are 2 main trends of consumption (i) regular OF consumers using standard distribution channels (supermarkets) and (ii) hardcore consumers adopting alternative channels (box delivery, farmers' market, specialty stores, and small grocery stores). According to Smithers et al. (2008) direct channels such as the farmer's market is targeted toward consumers that look to interact - socially - with the producers, ask them question about their production methods, food origin and variety, and cooking tips. On the other hand, conventional distribution channels, characterized by longer channels where consumers do not see and interact with the producer and where the information about food is limited, is targeted toward consumers that look for a one-stop grocery shopping experience. Distinct trends are thus observed in the organic food distribution. Each trend has its own development strategies but caters to consumers having different OF consumption motives, and base their choices on different sources of information and trust dimensions. Therefore, studying trust orientations regarding OF points of purchase along with trust toward brands, certification and labels, is very important especially knowing that the recent increase in OF consumption showed that it is strongly related to the consumers' trust in their food.

3.2 Certification and labeling: Building trust

Certification and labeling systems serve as tools to enhance distribution and market development, create trust, and foster confidence. It is a commitment from producers/farmers to work with certain standards of production. In 2009, the Canadian government implemented the Organic Products Regulation to regulate organic certification. In a nutshell, the new regulation requires mandatory certification for all agricultural products represented as organic in import, export and inter-provincial trade, or that has the federal organic agricultural product logo. This new certification logo has been recently created at the national level as a first step to standardize all certification processes across the country. According to Willer and Kilcher (2011), there are 80 countries using national standard of certification. The number of certified organic producers for the local market is growing and there are now Participatory Guarantee Systems (PGS) initiatives on all continents in terms of the number of farmers involved, with Latin America and India being the leaders. However consumers' confidence in certification standards in other countries and trust in their labels and products could be increased by the consolidation of standards and regulations between countries like Canada and the US, the world's first fully reciprocal agreement between regulated organic systems.

Issues of labeling and certification also still prevail, as many consumers are either unfamiliar with or confused by labeling due to lack of knowledge and their low ability to perform simple inference-making, leading to failure in decoding the information. Hence, consumers do not know to what degree they can trust certification labels. Nowadays, most countries have formulated standards for organic production and certification. This is considered as an important source of information about organic food quality and safety from the consumer's perspective (Hamzaoui and Zahaf, 2008). Just as branding of food products helps identifying the product to specific firms, organic labels are perceived as symbols of regulation, and therefore an important source of trust (Torjusen et al., 2004). There seems to be a need to make consumers trust both the product and any organism certifying this product. Public regulation and organic certification are traditionally a source of trust (Sassatelli and Scott, 2000), whereas a large number of private labels do not imply the same level of trust. Some countries use organic labels from different organizations as well as state labels. Switching to alternatives like adopting a single label at the national level (eg. in France) or regional level (e.g. in Europe) does not necessarily imply a better basis for label recognition and development of trust in these labels. With consumers wanting more in-depth information about the food and the food system, trust/mistrust in organic labels emerges as an important issue (Torjusen et al., 2004).

3.3 Country of origin and food mileage

Sustainable food systems represent one of the major innovations in the agricultural sector in the past decade (Thilmany et al., 2008). In fact, products labeled with credence attributes associated with local or organic food systems are enjoying high market penetration rates. However, food production and distribution patterns have undergone a major transformation. This has led to new market realities such as the importance of the country of origin and food mileage with regard to imported organic foods. For companies, ensuring sufficient supply volumes and supply continuity are becoming a major concern. Investing in developing countries is a mean to lock-in supply (Organic Monitor, 2006). In the case of Canada, imported organic products represent CA $252 million, of which 74% are from the U.S. The rest of imports is coming mostly from Chile, Mexico, China, Italy and Germany (Agriculture and Agri-Food Canada, 2008), with organic fresh vegetables and fruits being the largest imported categories (CA $223 million). From the consumer perspective, the origin of organic food possesses both predictive and confidence values (Luomala, 2007). This leads them to believe that they can make a reliable evaluation of food origin, and infer whether it is a good indicator for the desired product qualities, credible production control, and certification. Moreover, organic food imports also raise the issue of food mileage. This is directly linked to the sustainability of agriculture, as "organic food imports" do not match with local food production, freshness and community cohesion.

4. Demand side factors

Selecting food is one of the most common activities that consumers pursue many times each day. But this selection requires taking into account different goals (e.g. price and taste) and may involve a complicated decision-making process in order to satisfy these different goals. Different decisions with regard to organic food consumption will depend on internal and external factors affecting the decision process. Indeed, consumers might differ significantly with respect to use of and trust in information cues on organic food, knowledge and behaviour towards organic food, as well as socio-demographic profile.

4.1 The organic consumer profile

Organic food consumers have been profiled using a variety of variables such as purchase intentions or usage rate (cf. Davies et al., 1995; Fotopoulos and Krystallis, 2002a). The segmentation has also been based on demographic factors, food-related lifestyles, attitudes toward OF and purchase intentions, and frequency of purchase (cf. Brunso and Grunert, 1998; Brunso et al., 2004). Some common results on the socio-economic profile of organic food consumers show that organic purchasing grows as consumers reach their 30s and have no children. People who are among the highest spenders on OF are on average more affluent and younger (Padel and Foster, 2005). But lower income housholds also purchase organic food when convinced that organic food is better quality. Organic food consumers can be classified as "classic" or "emergent" consumers. The former is well-educated, a professional or white collar worker, willing to pay a premium for organics and to search out sources of organic food products (e.g. producer or farm markets). The latter is also well-educated, a professional, commited to personal health, and shopping in supermarkets as convenience is an important factor in his/her purchasing decision. Leger Marketing found in 2004 that out of 3.3 million regular and several time buyers of OF, 1% purchased on every food-shopping trip, 17% purchased them often, and 37% rarely purchased OF. Despite these results, Tutunjian (2004) notices that OF consumers share attitudes and values rather than demographics. The purchase of organic food products tends to be based on reasons ranging from dealing with food allergies to valuing the philosophy upon which organic farming is based. Overall, redefining OF consumers profile helps to better address the specific values underlying their food consumption.

4.2 Motivations to buy organic

Growing consumer demand for organic food (OF) has been attributed to consumers' concerns regarding nutrition, health, the environment, and the quality of their food (Fotopoulos and Kryskallis, 2002b; Larue et al., 2004; Shepherd et al., 2005). Further, various studies conducted in Europe and the US have explored the OF consumer behavior and have tackled the issue of determining consumers' motivations and preferences for organic products (Worner and Meier-Ploeger, 1999; Zanoli and Naspetti, 2002; Wier and Calverley, 2002; Yiridoe et al., 2005). Although some organic consumers are environmentally conscious, most studies confirm the predominance of egocentric values like health, attitude towards taste, and freshness that influence OF choice more than the attitudes towards environment and animal welfare (Millock et al., 2002; Fotopoulos and Kryskallis, 2002a; Zanoli and Naspetti, 2002). On the other hand, the main reasons that prevent consumers from buying OF are expensiveness, limited availability, unsatisfactory quality, lack of trust, lack of perceived value and misunderstanding of OF production processes (Fotopoulos and Krystallis, 2002a, 2002b; Verdurme et al., 2002; Larue et al., 2004). In Canada, consumers identify health, the environment, and support of local farmers as principal values explaining their OF consumption (Hamzaoui and Zahaf, 2008). These motivations and values are leading OF consumers to accept large price difference between organic and conventional food products.

4.3 Willingness to pay premiums

In the literature, a large body of research is dedicated to consumers' Willingness-To-Pay (WTP) for environmentally friendly products (Baltzer, 2003; Krystallis and Chryssohoidis, 2005; Laroche et al., 2001). This WTP appears to be a general tendency, and despite the increasing availability of organic food products, there are few studies that examined the variability of WTP for OF products in terms of product categories and OF consumers segments. Consumers' willingness to pay more for OF products reflects the "true" value of that product. This translates into price premiums, or the excess price paid over and above the "market" price (Rao and Bergen, 1992). Hence, the equation for marketers is very simple: no chemical pesticides, no chemical fertilizers, coupled with certification allow for a premium price strategy (Van Elzakker and Eyhorn, 2010). As stated by Vlosky et al. (1999), price and WTP a premium price are crucial elements of the OF consumers' behavior. In general, one reason why consumers are willing to pay a premium is to ensure product quality (Hamzaoui and Zahaf, 2009). But consumers differ in their level of willingness to choose higher-priced products (Krystallis and Chryssohoidis, 2005). We can expect regular consumers and hard core consumers to have different willingness to pay for OF products based on their respective motivations: health for regular consumers and environment, support of local community and health for the hard core consumers.

4.4 Trust orientations

To facilitate decision making in complex food markets, trust is an essential element. In general terms, when related to food, trust is seen as "an expression of the alternative to have to make an individual decision, and just assume that food is safe" (Green *et al.*, 2005; p.525). More specifically, there are particular information sources and organizations that are trusted to either provide safe food or to provide trustworthy information about that food. Considering the risks associated with product consumption, consumers will search for and adopt several risk reduction strategies (Mitchell and McGolrick, 1996; Brunel, 2003) such as brand image (Gurviez, 1999; Gurviez and Korchia, 2002), store image, or label references. These are all means to built trust in the product. Studying OF consumption, Sirieix et al. (2004) highlights two sets of trust orientations defined as indicators that consumers rely on in order to "trust": trust oriented toward several quality indicators, and trust oriented toward individuals. Therefore, trust can be oriented toward the brand, the label, but also toward partners like producers. Trust has been identified as an important strategic variable in the food industry (Bahr et al., 2004). Studying trust orientations is hence important to clarify the market position of organic products, sales channels and certification authorities. In fact, increasing OF consumption seems to be directly linked to consumers' trust orientations and values. Hamzaoui and Zahaf (2008, 2009) highlighted in their study Canadian consumers' concern about quality indicators of OF such as trust in the certification label, trust in the product's country origin, but also trust in the type of channels of distribution used.

5. Objectives

In order to target more efficiently consumers, we need to provide a more precise and useful profile of organic food consumers, who they are, what they eat, how they buy, where they buy, and why they eat organic. This will lead to an in-depth understanding of the organic food industry, the major forces shaping it, and the current market structure, as well as an understanding of the challenges faced by the main players of the organic food industry. Moreover, it will provide a detailed assessment of the actual situation in the OF distribution system, i.e., superstores, specialty stores, and farmers' market. This will help to understand the importance of the value delivery network in creating value added to the OF supply. Hence, our objectives are:

i. Assessing the importance of the channels of distribution, labeling and certification process and food mileage in the organic food market.
ii. Determining OF consumers' purchasing behaviour in terms of how OF consumers buy, where they buy, their sources of information, their trust orientations, and the trusted channels of distribution.
iii. Clustering OF consumers with regards to their psychographics.

6. Design and procedure

6.1 Design

To address the abovementioned objectives, a mixed design is needed. On one hand, we need to assess the supply side situation by conducting personal in-depth interviews with organic food producers, channel intermediaries, final retailers and certification bodies. On the other hand, we need to survey organic food consumers to assess their consumption behavior/patterns. For the supply side, in-depth interviews were conducted with store managers of superstores, specialty stores, and farmers' markets (producers) in Thunder Bay, Toronto, Ottawa, and Montreal. Interviews were based on an interview guide and lasted about 45 minutes to 1 hour. The guide probes various channels members, distributors, and producers of OF to discuss the actual structure of their distribution channel, their marketing strategies, and trust issues related to their distribution strategies. The interviews were recorded, transcribed, coded, and analyzed by the researchers using content analysis (cf. Kassarjian, 1977). Two separate judges coded the data.

For the demand side, a survey was administered to consumers in Thunder Bay and Ottawa. The population targeted for this study is OF shoppers. For purpose of gaining a good representation, respondents needed to fit within a specific profile. The idea was to select randomly organic food consumers that make their purchase mainly in specialty stores, grocery chains or local markets. Data was collected using two administration modes: in-person and online (using coupons with the survey URL). This helped to balance the proportion of consumers shopping in different channels of distribution.

6.2 Distribution interview guides

The first objective of this study was implemented using a qualitative design. A total of 42 in-depth interviews were conducted in Winter 2011 in different Canadian cities (including French and English speaking provinces). The objective is to determine and understand current and new trends in the organic food industry, the distributors' perceptions of consumers' concerns and level trust in organic food, and finally, how consumers' concerns are addressed. The interview guide is composed of three main sections. The first section deals with the structure of the channel of distribution while the second and third sections deal with how suppliers perceive consumers' demand and concerns. This three-prong interview guide helps to determine how distributors/suppliers manage similarities and differences between what consumers want and what they offer them. Distributors were profiled as follow: (i) by channel size and type, (ii) by organic food products variety, and (iii) by channel position (retailer, wholesaler, etc.). All interviews were analyzed using content analysis (cf. Kassarjian, 1977).

6.3 Consumer survey measurements and scaling

To test the abovementioned 2nd and 3rd objectives, a structured questionnaire was designed to gather data that measure the variables used in this research. Prior to administering the survey, a pre-test was done and minor modifications were made. The questionnaire is structured into three sections. The first section deals with consumers' general opinion about organic food, consumption and shopping habits, and last, reasons for buying organic (measured on a 5 point Likert scale). The second section of the survey deals with trust dimensions (measured on a 5 point Likert scale). Finally, the third section is structured to design a socio-demographic profile of our respondents. Most of the questions in the survey were adapted from Sirieix et al. (2004), and Fotopoulos and Krystallis (2002). A total of 350 questionnaires were collected, and 324 questionnaires were usable. Data was cleaned and missing values were replaced using the mean. All variables were tested to check their internal consistency. Reliability tests were coupled to a series of factor analyses to determine the structure of the data. Factor analysis also helps to test if the items are measuring the right constructs. Results showed that Cronbach alphas were in the range of 0.727 to 0.850, which is good for an exploratory study (Hair et al., 2006). All variables except "trust" have a unidimensional structure with factors loading ranging from 0.583 to 0.893. Three dimensions were found for "Trust": (i) brand and store trust: trust in the brand and the store where the purchase is made, (ii) prior experiences: all information related to prior experiences with the product and involved in building trust with regards to trusted labels, brands, and points of purchase, and (iii) organic labels trust: unknown factors such as lack of credibility of the organic labels, meaning of "organic", and lack of trust in the quality stated in the organic labels. Hence, the trust scale has been split into three dimensions.

7. Supply side analysis: Qualitative study

In order to get a representative image of what the organic food distribution system looks like, several and various players in the organic food distribution channels have been interviewed. This includes: producers, farmers, store managers, distributors, wholesalers, and certification bodies representatives. This gives also a wider perspective on the structure of the organic food industry, the new trends in the organic food market, and the challenges faced by all channel members. Table 1 shows all themes generated from the interviews.

7.1 OF Industry and market

The first theme is related to the structure of the organic food industry. The guide probes the interviewee to describe the current situation of his/her distribution organization system, the organic food market and its negative/positive aspects. With this regards all channel members as well as producers see a big potential for this industry. From their perspective, consumers seem to be attracted by the healthy aspect and nutritional value provided by organic foods, while corporations and distributors are attracted by the profitability of this growing market. Further, consumers are becoming more educated about organic foods and

THEMES	SUB-THEMES
Current OF market	Unstable supply/availability Pricing Supply Driven by demand Consumer's education Better quality
OF brand growing	Positive branding More choice for the consumer
Trends	Local food Environmentally friendly Gluten free produce New product lines
Distributor's Channel	Depends on the direct last channel members Differentiation Get most information about the market from the delivery companies Grocery chains vs specialty stores
Feedback/Expectations	Bottom-up communication
Demand	Education Growing
Changes in OF consumers	Definition of organic Smarter consumers
Reasons to buy	Health Environment Taste
Concerns	Not in store
Types of consumers based on trust	Trust labels/not trust the labels Clear differences Brand name
Increase trust	Pricing accuracy Knowledge of clerks / advices Educating consumers Quality Knowing the producer
Sustainable development	Very important Not dangerous Will be a condition to access the market A way of differentiation Could be used to increase trust Competitive advantage

Table 1. List of Generated Themes

they are asking for more organic products, *"I think, with more and more research, healthy eating takes place. Consumers are becoming more educated and with that are making better decisions for themselves and for their loved ones"*. This new type of consumer is willing to pay high premiums, up to 100% (cf. Hamzaoui and Zahaf, 2009). Since OF are premium priced products then the *expected* performance of these products is higher then regular products. Having said this, there is clear-cut in the respondents' perception of the OF industry and market. The shorter the channel of distribution (sales at farms gate, direct from producer/farmer), the simpler the logistics. The trust relationship between consumers and these sellers is very strong and is based on direct knowledge of how the product is grown, how it is certified, and marketed: *"trust can be increased by advice, and more importantly the relationship established with the consumer"*; and more importantly it is based on the direct impacts on the local economy and the environment. Food mileage and local foods are the most important new trends described by channels members: *"They would like organic local... they also go for local because they know where it's from"*. Consumers using short channels look for different product attributes and have different motivations than consumers using standard channels of distribution, eg. grocery chains.

The OF market is growing and new product lines with additional attributes are emerging: organic products that are local or fair trade. Organic is becoming omnipresent in the food market. It targets all consumers as well as actual regular OF consumers and hardcore consumers. Because of their motivations and ethical values, they are interested in fair trade organic products or local organic products. These differences lead to a distribution system based on different channels: organic foods are mainly sold through standard channels of distribution, whereas fair trade and local organic food require shorter channels. Weekly baskets delivered to hardcore consumers seem to be the best way to satisfy these consumers looking for ethical organic products: *"a basket delivered weekly is a key to have products that are organic and local"*. It is clear that the OF supply is driven by the new trends in the OF market.

7.2 Current distribution system

The second theme deals with the channel members' perception of the industry, the distribution system, and their distribution structure. Results are consistent with the first section of the interviews. All channel members from various channels agree to say that consumers are becoming more educated and make smarter food choices. However, there are clear differences in their purchasing behavior. Shorter channels managers stated that consumers buying at their point of sale have certain needs and certain motivations to buy organic foods. Conversely consumers buying from longer channels are looking for a different shopping and consumption experience. This is directly related to the OF adoption process. Consumers trusting the labels and certifications are either in the interest-evaluation-trial phase while consumers trusting stores are in the adoption phase. In fact, there are different levels of trust according to the channel members: trust related to the labeling and certification, trust related to the channel of distribution, and trust of the producer.

7.3 Sustainability

The third theme deals with (i) the importance attributed to sustainable development in the OF industry, (ii) if sustainable development may be considered as a competitive advantage in relation to organic food, and (iii) if it can increase consumers' trust in organic food products. Generally speaking, there is some consensus among the interviewed suppliers on these three aspects. The importance attributed so far to sustainable development is not really strong with few exceptions. Adding the sustainability claim to organic seems difficult to justify as organic still has a long way to go, especially for regular and emergent OF consumers: *"You need to educate before going with new ideas and you need to better develop consumers' knowledge in organic before"*. Respondents from specialty stores as well as supermarkets agree that this "sustainability" aspect of the product is not targeting every consumer that wants to eat healthy and is still learning about organic. Some respondents consider that sustainability can constitute a competitive advantage if the targeted consumers are the ones already caring for these characteristics, in other words hardcore consumers that highly value the environment. Specialty stores representatives highlight the importance of the origin of the product more than the general concept of sustainability: *"where it comes from is more important"... "having local products and knowing the origin is the most important thing for organic"*. Skepticism is what emerges from the interviews as (1) most respondents highlight that it will bring confusion because consumers will get too much information on top of all what is related to organic, and (2) it is not sure as how much it will add to organic and how organic will be defined within these new attributes and claims. Indeed, when confronted with too much information, there is a risk of information overload and potential adverse effects because of consumer misunderstanding.

8. Demand side analysis: Quantitative study

8.1 General profile of organic food consumers

Our sample is composed of 324 consumers. In a first step, consumers are classified as follow: if respondents buy organic at most once a month then they are classified as non-regular organic food consumers (non-RC) while if they consume organic food very often then they are tagged as regular organic food consumers (RC). This is a basic grouping method (Cunningham, 2001). Accordingly, respondents are distributed as follow: 62% of RC and 38% of non-RC. Further, the typical profile of our respondents is: female (69.7%); aged 25 to 35 years old (49.1%); single (34%) or married (37.7%); household composed of 2 to 3 persons (47.8%); have at least an undergraduate degree (69.9%); works as a professional (26.9%) or is white collar (22.8%); buys at least 2 organic food products (90.8%); eats mainly organic fruits and/or organic vegetables (24.1%); buys organic food mainly from supermarkets (31.2%); spends on average $100 in organic groceries (58.4%); considers nutritional value, freshness, healthiness, and taste as the major factors for buying organic food product; and finally is happy with his/her organic consumption experience (90.8%). Cross-tabulations with Chi-square testing were used to explore the relationships between the main different indicators. Results show that age as well as monthly spending, satisfaction, and OF product category are good predictors of the type of consumer. 35% of non-RC spend at most $100 in OF groceries while only 25.1% of RC spend the same amount. Further, 31.4% of RC have a monthly spending in OF groceries of $100 to $400. This is explained in part by the type of

Product Category	Non-RC	RC
Dairy	8.3%	41.9%
Fruit	27%	59.7%
Bread	11.1%	46%
Meat	7%	27.9%
Vegetables	28.3%	57.5%
Prepared food	8.6%	36.5%

Table 2. Purchase Distribution for RC and non-RC per Food Category

products bought by these consumers (cf. Table 2 for more details). All OF categories are evenly represented for RC while most non-RC buy mainly fruits (27%) and vegetables (28.3%).

8.2 Trust orientations

There are three dimensions of trust considered in this research: (i) trust towards the brand/store, (ii) prior experiences with OF and (iii) organic labels trust. We ran a three t-test for independent samples to uncover differences/similarities between the RC and non-RC on the trust dimensions. The results are conclusive for the organic labels trust, and brand and store trust (cf. Table 3). This means that RC and non-RC perceive differently the organic labels trust in terms of credibility of organic labels, meaning of "organic", and lack of trust in the organic label claims. Moreover, non-RC show a higher degree of uncertainty on all trust items in comparison to RC. This was somehow expected as non-RC are still unsure about what organic is.

Trust Dimensions	Consumers	Mean	t-test Significance
Brand and Store	RC	3.533	0.138
	Non-RC	3.415	
Organic labels trust	RC	2.643	0.000*
	Non-RC	3.116	
Prior Experiences	RC	3.280	0.219
	Non-RC	3.195	

Table 3. T-tests for Levels of Trust *sig. at 5%

Moving to trust towards the brands and stores, we notice that even though there isn't any significant difference between RC and non-RC, both types of consumers score high (5 point Likert scale). This means that RC and non-RC trust the brands and the store from where they buy OF products. In addition, it is important to note that all consumers score high on the trust dimension related to prior experiences. Hence, RC and non-RC have the same - relatively high - level of trust when it comes to their prior experiences with the store and organic labels. This is directly related to consumers' loyalty and their habitual purchase pattern, i.e., a consumer wants to use the same point of sale and same product/organic label if they are satisfied with their purchase. In order to test this, we explore the relationship between consumers' satisfaction and their prior experience for building trust. The chi-square test was conclusive (sig. = 0.000 < 5%) showing that satisfaction determines the type of consumers, i.e., RC or non-RC. Most of the RC are satisfied (38.3%) or very satisfied (59.7%) with their previous OF consumption while 52.6% of non-RC are satisfied with their previous OF consumption (only 25.4% of them are very satisfied). This difference might be based on the fact that non-RC do not exactly know what to base their satisfaction on as they are not as well educated as RC about organic food. Lastly, consumers consider that family, friends, and/or scientific articles are the most important sources of information on OF. Conversely, when asked to rank the most trusted sources of information, consumers listed small shops and consumer organizations as being the most reliable sources of information.

8.3 Channels of distribution

In order to understand where OF consumers buy and why they prefer some channels of distribution over others, we need to determine the most used channels of distribution. We first tested the relationship between the type of consumer (*RC* versus *non-RC*) and most used channels of distribution using cross-tabs and Chi-square tests. Then we tested the association between trust orientations and choice of channels of distribution using t-tests (*channel users* versus *channel non-users*). This helps to understand who the users are: RC or non-RC; and their trust orientations.

8.3.1 Channel user/non-users vs OF RC/non-RC

Results from the cross-tabulations give the following distribution of OF consumers. It is clear that most RC buy from short channels while non-RC buy from standard channels. A more in-depth analysis of the most used channels and the most trusted channels is needed.

Channel of Distribution	Non-RC	RC
Supermarket	21.5%	17.6%
OF store	5.2%	16.9%
Directly from producer	3.6%	5.2%
Local market	1.6%	7.8%
Convenience store	1%	0.3%
Health food store	1.3%	11.1%
Home delivery	0%	2.6%

Table 4. Distribution of RC/non-RC by Channel of Distribution

8.3.2 Most used channels of distribution

A descriptive analysis of the data shows that 74.6% of all respondents buy at least from two different outlets. Further, the *most used* channel of distribution is the supermarket (31.22%), followed by the organic food stores (27.77%), and local markets (27.17%). As expected, RC consumers represents the largest proportion of channels' users for all channels of distribution. In order to test if trust orientations are associated with the choice of the channel of distribution, t-tests were run. Results are summarized in Table 5.

Table 5 shows that channel users and non-users of specialty stores and supermarkets have different trust orientations, i.e., prior experiences, organic labels trust, and brand and store trust. Conversely, there is no difference between the level of trust of users and non-users of local markets. Direct channel users and non-users have different levels of organic labels trust and prior experiences. This makes sense since OF consumers use direct channels - represented by farmers or organic producers - if they know who the producer is and what the products are. The purchase situation and framework in this case are very context specific, and hence consumers' trust orientation is based on a direct relationship with the producer. As far as the supermarkets go, the only important trust orientation is the one related to the store and the brand, which is in accordance with Sirieix *et al.* (2004) findings. The crux of shopping at supermarkets is that these channels of distribution are the most used point of sale because of their convenience, and may not be necessarily trusted. Further, Chi-square tests show that the only significant relationship is between on one hand, organic food stores users and non-users, and on the other hand the type of consumers, RC and non-RC. This means that the type of channel of distribution, and its users, does not determine the type of consumers (RC or non-RC) of that channel of distribution. To recapitulate, consumers' trust orientations determine their choice of the *most used* channel of distribution, and channels adoption is not related to the type of consumers.

	Trust Dimensions	Channel users	Mean	t-test Significance
Direct Channel	Prior Experiences	Users	3.102	0.014*
		Non-users	3.314	
	Organic labels trust	Users	2.682	0.048*
		Non-users	2.866	
	Brand and Store	Users	3.590	0.081
		Non-users	3.446	
Local Market	Prior Experiences	Users	3.270	0.522
		Non-users	3.218	
	Organic labels trust	Users	2.850	0.439
		Non-users	2.783	
	Brand and Store	Users	3.462	0.381
		Non-users	3.532	
Organic Food Stores	Prior Experiences	Users	3.354	0.001*
		Non-users	3.072	
	Organic labels trust	Users	2.707	0.001*
		Non-users	2.987	
	Brand and Store	Users	3.601	0.002*
		Non-users	3.306	
Supermarkets	Prior Experiences	Users	3.257	0.787
		Non-users	3.233	
	Organic labels trust	Users	2.804	0.843
		Non-users	2.822	
	Brand and Store	Users	3.425	0.008*
		Non-users	3.363	

Table 5. T-tests for Most Used Channels of Distribution *sig. at 5%

8.3.3 Trusted channels of distribution

The most *trusted* channel of distribution is the organic food store followed by health food stores, and the direct channel producer-to-consumer. Interestingly enough, supermarkets ranked fourth ahead of local markets (all means higher than 3 on a 5-point Likert scale). Data has been recoded to address some of the complex issues related to distribution. Respondents were asked to assess seven OF channels of distribution using a 5-point Likert scale ranging from 1 = strongly disagree to 5 = totally agree. Each scale has been recoded as follow: high trust consumers (score 4 to 5) and low trust consumers (score 1 to 3). 12 t-tests have been run to determine if there is a significant difference between the level of trust and the trust dimensions (store and brand, organic labels trust, prior experiences) of the 4 major channels of distribution (organic food stores, supermarkets, direct channel, and health food stores). Results show that consumers trust all channels of distribution to certain degrees, but also that there are no major differences between RC and non-RC for all levels of trust and most channels of distribution. Consumers trust all channels of distribution but consumers' trust orientations intervene only when choosing to shop from an organic food store. Further, when it comes to the most used channel of distribution, it appears that trust orientations are the main cause for channels of distribution utilization.

8.4 OF segments

8.4.1 Profiling OF segments

A combination of usage rate, trusted points of purchase, and consumer's lifestyle has been used to cluster OF consumers (cf. Hartman and New Hope, 1997). In this study, all aspects related to community-supported agriculture, support for the local economy and cooperative growers, food mileage, as well as environmentally friendliness concerns will be referred to as consumers' lifestyle. This provides a more realistic idea of the OF market segments.

Segments	Acronym	Description	Percentage	Mean Range
True OF consumers	TOF	RC, high trust, and principle oriented	52.3%	3.27 to 4.54
Occasional OF consumers	SOF	Non-RC, high trust, and moderately principle oriented	36.4%	2.54 to 3.75
Inexperienced OF consumers	IOF	RC, low trust, principle oriented	11.3%	0.32 to 4.36

Table 6. Cluster Analysis for a Combination of the Variables

	TOF	SOF	IOF	Chi-Square
Gender	All segments are mainly represented by females			0.072
Age	Young consumers (25-45)	Young consumers (18-35)	Young and older consumers (25-35 and 55+)	0.000*
Marital status	Single-married-divorced	Single-married	Single-married	0.000*
Income	50-70K	30-50K	0-15K 70-100K	0.003*
Place of residence	Mainly downtown			0.001*
Level of education	University degree (undergraduate as well as graduate degrees)			0.000*
Occupation	Professionals	Professionals and white collar	Professionals, white collar, and students	0.000*
Monthly spending	81.6% spend $50-400	61.8% spend $0-50	23.5% spend $0-50 19.7% spend $100-200	0.000*

Table 7. Socio-demographics of OF Consumers and Tests of Association *sig. at 5%

Cluster 1 is composed of *true OF* consumers, or TOF, as they buy OF products frequently, trust almost all channels of distributions, and are principle oriented. Further, there are some variations in their trust levels per channel of distribution as they show moderate trust with regards to supermarkets (mean of 3.44 on a 5-point Likert scale), marketplaces (3.36), convenience stores (1.94), and home delivery (3.27). *Sporadic OF* consumers, or SOF, are consumers who do not buy OF on a regular base. They trust all channels of distribution but have neutral attitudes toward supporting the local economy and the environmental friendliness of OF products. Conversely, *inexperienced OF* consumers, or IOF, are consumers who consume OF products on a regular base but do not trust any channel of distribution, as they don't feel confident when buying OF products in those points of purchase. However, they are principle-oriented consumers. It is important to note that that TOF have a higher frequency of purchase than IOF.

TOF, SOF, and IOF consumers have been profiled using the socio-demographics data. Table 7 depicts all socio-demographic characteristics of the 3 segments. We used relative measures, as there were missing values. It is readily seen that age, marital status, income, place of residence, monthly spending, level of education, and occupation are determinant of the OF clusters. However, results show also that gender is not a determinant of the segments (Chi-square test is not conclusive: sig = 0.072 > 5%).

8.4.2 Motivations

Consumers were asked to rate their reasons to buy OF. Generally speaking, they value health, taste, environmental friendliness, superior quality and the support of the local economy. A one-way ANOVA was run to test differences between the three segments (TOF, SOF, and IOF) with regards to the five reasons to buy OF. Overall, consumers in the three defined segments have different reasons to buy OF (cf. Table 8). It is also clear that TOF have the highest scores on all reasons to buy, with health, environment and local economy being the most important ones. Conversely and as expected, IOF score the lowest on all reasons to buy.

Reasons to Buy	TOF	SOF	IOF	Sig.
Health	4.72	4.35	3.95	0.000*
Taste	4.39	3.86	3.52	0.000*
Environment	4.69	4.35	3.94	0.000*
Quality	4.26	3.66	3.49	0.000*
Local economy	4.68	3.93	3.55	0.000*

Table 8. OF clusters and Reasons to Buy *sig. at 5%

8.4.3 Predicting consumers' membership

In order to assess the predictive power of each variable in predicting cluster memberships, a discriminant function analysis was run. This analysis is used to determine which variable(s) discriminate between two or more naturally occurring groups. For instance, after segmenting the market using cluster analysis, managers would like to know (i) how to classify new OF costumers according to a set of variables, and (ii) what variables allow the best allocation and targeting of OF consumers. This is achieved through discriminant analysis techniques. Results from Table 9 show that the frequency of OF purchases is the best predictive variable as it correctly classifies 71.2% of the respondents. It is clear that consumers in the three segments, i.e., TOF, IOF, and NOF, have a very complex psychographic profile. This should lead companies to use an optimal mix of variables. In our case, the combination of all variables provides 91.9% predictive power meaning that 9 consumers out of 10 are correctly classified if we consider a combination of the variables listed in Table 9.

	Classification accuracy	Discriminant functions	Wilks' Lambda	Significance
Reasons to buy OF	51.0%	1	0.818	0.000*
Trust (brand, label, store)	48.5%	2	0.835 0.965	0.000* 0.005*
Support local economy	66.9%	1	0.777	0.000*
Environmentally friendly	64.9%	1	0.695	0.000*
Most used point of purchase	58.6%	1	0.377	0.000*
General attitudes	54.6%	1	0.809	0.000*
Sources of information	39.7%	1	0.903	0.000*
Trusted sources of information	33.9%	1	0.949	0.000*
Frequency of purchases	71.2%	1	0.587	0.000*
Combining all variables	91.9%	2	0.124 0.396	0.000*

Table 9. Discriminant Analysis for TOF, IOF, NOF

9. Discussion and managerial implications

The objective of the qualitative and quantitative studies conducted with various players of the organic food market was to gain a wider perspective on the structure of the organic food industry. This allowed us to uncover new trends in the organic food market and to assess the challenges faced by the organic food industry and its players.

9.1 OF market growth and supply management

All the players from the supply side interviewed for this study generally agreed that the market for organic products is growing and shows that it has a long life and substantial growth opportunities. More specifically they mentioned an increasing diversification of products and distribution channels. As mentioned by Zander et al. (2007) 'conventional' entrepreneurs and corporations have been attracted by the remarkable success story of the organic sector. The increasing number of distribution channels is mainly based on more supermarkets and food store chains offering OF products and widening their offer of organic food at more competitive prices. This is in accordance with Dimitri and Oberholtzer (2009) mentioning that the recent burgeoning of organic store brand products has contributed significantly to increase organic sales. Most conventional channels are increasing their organic sales using traditional marketing strategies for organic food, including organic versions of conventional brands. This is done to satisfy the needs of a wider number of OF segments.

From the producers and farmers perspective, being able to expand supply is a big issue that translates into poor supply reliability and poor availability at the demand level. Indeed, with the growth in popularity of organic food products, more wholesalers have entered the organic food supply chain. They have been encouraged by chain stores that want to work through them because demand is up and they need larger quantities at regular delivery times. Consequently, imports from regions (eg. California) with large organic farming activities still prevails. This rises the question of what organic exactly is, if it is perceived as limited to local or if imports can also be considered as organic. Having different channels of distribution somehow addresses the different perceptions of the OF market segments on these matters. Farmers, local markets and specialty stores emphasize the origin of the organic products and provide complete information to OF consumers.

For supermarkets, the diversification of the offer is the main driver of the market growth. According to Padel (2005) the relationship between the organic suppliers and the conventional retailers in the mainstream food chain is a relationship between small and big volume actors. But for organic suppliers, the most reasonable access to consumers in many countries remains the conventional retailers (Bahr et al., 2004), highlighting an important need for closer collaboration between the supply chain actors.

From the organic food specialty stores' perspective (independent stores as well as chain stores), the organic market shows differences between supermarkets and specialty stores in terms of variety, price and quality. In other words, supermarkets are able to provide consumers with a larger variety, lower prices and convenience whereas specialty stores differentiate themselves with the quality and the origin of their products. As for producers, the main difference between suppliers is established in terms of short-direct / long channel of distribution, with producers offering traceability and quality.

9.2 Organic food consumers

Suppliers also provided their perceptions on several organic consumers' characteristics. For most suppliers, consumers are in general more knowledgeable and are looking for authentic products, health, quality, and taste. Their level of knowledge as well as their motivation to consume organic products seems to differ depending on the point of sale they mostly use. In other words, consumers buying from producers/farmers are clearly looking for proximity with the producer, fresh products and quality, and a better understanding of the organic process and show a clear interest for its impact on health and the environment. Consumers mainly using standard channels of distribution are looking for convenience, healthy products and taste. These consumers do not have a high knowledge of what organic is and seem to get confused between organic and natural products. Organic specialty stores describe their consumers as more knowledgeable and looking for health, quality and taste.

Trust is also of extreme importance in the organic food networks as its added value is mainly based on the production methods.

Several organic labels are present on the market. This induces some confusion, as consumers do not know which one(s) to trust. Therefore, the organic labels that should play a central role do not seem to have achieved that position in the OF consumers' decision-making process yet. Overall, distribution channels link consumers' trust in OF to different dimensions: organic labels, brands, traceability, advice, and/or store reputation. Because of the differences in these trust dimensions, providing standard information for all OF consumers may not be the best communication strategy based on consumers' specific interests and knowledge. For consumers' mainly purchasing their OF products in supermarkets, organic labels is mainly what they trust as well as organic brands (mainly store brands). This is in accordance with Sirieix et al. (2009) showing that OF consumers buying in supermarkets rely on organic labels and brands. Consumers purchasing in specialty store trust the store itself, the sales persons' advices, the products' traceability (transparency of the supply chain) and to a much lesser extent some organic labels they know. Hence, communication on the products' quality and traceability, advices and information provided by store managers and sales persons (and store reputation) could increase consumers' trust in OF. For consumers purchasing from producers and farm markets, traceability is the main element of trust. Further, close and direct communication and relations between farmers and consumers would help maintaining and/or increasing trust as well as feedback on product quality and taste. In accordance with our results in terms of most used and trusted sources of information, these consumers subscribe to health and wellness magazines and read books on the subject (Zepeda and Deal, 2009). Last, brands are not a major factor that Canadian consumers can yet base their trust on. Finally, the issue of certification labels and related trust arises for the majority of consumers when considering imports of organic products, as other countries are indeed perceived as not meeting the same organic certification requirements as Canada. Imports also raise the issues of product quality, traceability and food mileage.

9.3 Growth perspectives and sustainability

Based on the supply side interviews, growth perspectives are directly linked to the challenges expressed by the respondents: maintaining and increasing consumers' trust in OF and in their distribution systems, and adapting their offer to new trends in OF demand. Results from the consumer survey also highlight two main new trends in consumer demand: food mileage and local foods. These new trends are more noticeable for OF consumers making their purchases in specialty stores and local markets rather than in supermarkets.

While sales of organically certified products have grown, the sector has to face new market entrants making green and ethical claims. In other words, the organic sector faces the challenge of an increasing number of other standards and brands competing for green and ethical segments in the consumer market. Zander and Hamm (2010) highlighted the interest of some OF consumers in several ethical concerns such as animal welfare, preservation of biodiversity, fair prices for farmers and local supply chains. But as shown by our results, only some ethical concerns are being actually considered by OF consumers, mainly hardcore consumers. Further, organic food is not yet frequently associated in regular consumers' mind with ecological and social sustainability.

The regular consumer is committed to personal health and being health conscious, sees the absence of chemical and pesticides on organic food as a reputation for quality. The hardcore consumer is committed to the environment, but is also concerned by the evolution of "industrial organic" and sees the locally grown foods and the development of local food systems as an alternative for a more sustainable food system. A new term has even emerged, "beyond organic", to describe the importance of other qualities than the ones defining what organic is. Food mileage, reliance on local resources and environmentally sustainable food production on top of health concerns are what makes the added value of these "beyond organic" products for hardcore consumers.

To recapitulate, the OF market is growing and new product lines are emerging and being marketed. We are moving slowly from a situation of exclusivity to a situation where all consumers in the market are targeted. Hardcore consumers are looking for "fair trade organic foods", while regular OF consumers look for regular organic foods or at most local organic foods. This is very important to know by marketers and decision makers as it determines the real motives and reasons to buy OF and where to buy OF. Fair trade OF and local OF require very short channels of distribution; final prices are high; and the target market is small. Conversely, regular organic foods do not require short channels and are sold using standard channels of distribution. The direct consequence is a fast growing and profitable market segment using standard marketing tools.

10. Conclusion

Consumers' interest in organic food has exhibited continued growth for the past two decades, which has attracted entrepreneurs and corporations seeing a big potential for this industry. This led to the creation of standards and regulations to guide the OF industry. There are clear challenges on both demand and supply sides. Consumers are becoming more sophisticated in their purchasing decisions of OF, and companies are focusing on supply chain management in order to ensure high quality, traceability, and supply continuity. The OF industry also faces some other challenges: (i) maintaining and increasing consumers' trust in the OF products and the OF industry in general, and (ii) facing competition from other sustainability labels and initiatives. The OF industry and all its stakeholders will have to elaborate strategic responses to these opportunities and challenges. The results provide an insight into the structure of the organic food industry based on studies conducted with suppliers and consumers of OF products. The increasing number of OF consumers and the changes in organic product retailing still leads to an important imbalance between supply and demand high operating costs as well as poor supply reliability.

Our results also show that there are 3 types of OF consumers based usage rate, trusted points of purchase, and support for the local economy and the environment. True organic food consumers, or TOF, buy OF products frequently, trust almost all channels of distributions, and are principle oriented. Conversely, *sporadic OF* consumers, or SOF, are consumers who do not buy OF on a regular base. They trust all channels of distribution but have neutral attitudes toward supporting the local economy and the environmental friendliness of OF products. Lastly, *inexperienced OF* consumers, or IOF, are consumers who consume OF products on a regular base but do not trust any channel of distribution in particular but are principle-oriented consumers. All these market segments have different consumption preferences and hence, trust. They use differently the existing channels of distribution. When comparing the channels of distribution, it clearly appears that consumers buying from short channels have specific motivations to buy organic foods that differ from consumers buying from longer channels . This is directly related to the OF adoption process. It is also important to note that when it comes to the most used channel of distribution, trust orientations are the main cause of channels of distribution use. The most *trusted* channel of distribution across all OF consumers is the organic food store, then health food stores, and producer-to-consumer channel. When it comes to trust in organic labels, RC and non-RC have different views in terms of credibility of organic labels, meaning of "organic", and lack of trust in the organic label claims. Moreover, non-RC show a higher degree of uncertainty on all trust items in comparison to RC. Lastly, the organic claim itself still has a long way to go and adding the sustainability benefit can only bring confusion to most OF consumers. Having said this, suppliers clearly state that today what appears to be an important attribute for an increasing number of OF consumers is the local origin of the organic product. Whereas purchasing local foods is a possible pathway for achieving sustainability, it is not yet perceived as the real mean to engage in consumption practices that lead to a more sustainable food and agricultural system.

The following supply-demand model is based on all the results of this study and summarizes the major findings. It is clear that the model is a first step towards understanding the OF market dynamics from the supply and demand sides. The model is simple but depicts the major trends and logistics in the industry. More works needs to be done in terms of having more exhaustive samples that allow to strengthen the model structure and render it more robust.

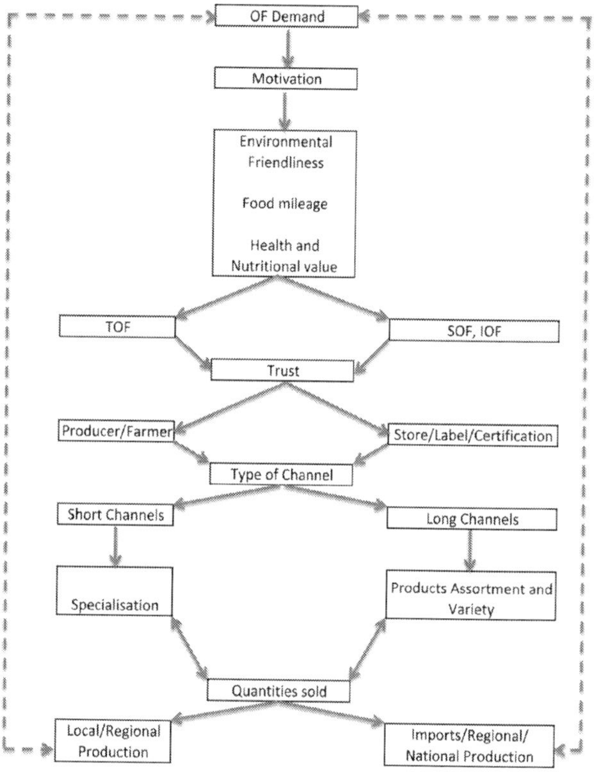

Fig. 1. Organic Food Demand-Supply Model

11. References

Agrifood Canada, Organic Production, http://www4.agr.gc.ca/AAFC AAC/display-afficher.do?id=1183748510661&lang=eng.

Bahr, M., Botschen, M., Laberentz, H., Naspetti, S., Thelen, E., and Zanoli, R. (2004), *The European consumer and organic food*, School of Management and Business, University of Wales Aberystwyth, Aberystwyth.

Baltzer, K. (2003), "Estimating willingness to pay for food quality and safety from actual consumer behaviour", paper presented at the 83rd EAAE Seminar, Chania, 4-6 September, available at: www.maich/eaae.gr

Bean, M., and Sharp, J. (2011), "Profiling alternative food system supporters: The personal and social basis of local and organic food support", Renewable Agriculture and Food Systems, Vol. 26, 243-254.

Brunel, O. (2003), "Les strategies d'ajustement au risqué alimentaire", Proceedings of the FMA International Congress, Gammarth, Tunisia.

Bruntland, G. (1987), "Our common future: The World Commission on Environment and Development", Oxford, Oxford University Press.

Brunso, K., and Grunert, K.G. (1998), "Cross-cultural similarities and differences in shopping for food", Journal of Business Research, Vol. 42 (3), 145-150.

Brunso, K., Scholderer, J., and Grunert, K.G. (2004), "Closing the gap between values and behavior - a means-end theory of lifestyle", Journal of Business Research, Vol. 57 (6), 665-670.

Chryssohoidis, G.M. and Krystallis, A. (2005), "Organic consumers' personal value research: testing and validating the list of values (LOV) scale implementing a value-based segmentation task", Food Quality and Preference, Vol. 16, 585-99.

Davis, A., Titterington, A.J. and Cochrane, C. (1995), "Who buys organic food? A profile of the purchasers of organic food in Northern Ireland", British Food Journal, Vol. 97 (1), 17-23.

Dimitri, C., and Oberholtzer, L. (2009), "Expanding demand for organic foods brings changes in marketing", Markets and Trade, March, 3.

Fotopoulos, C., & Krystallis, A. (2002a), "Purchasing motives and profile of the Greek organic consumer: a countrywide survey", British Food Journal, Vol. 104 (9), 730-764.

Fotopoulos, C., & Krystallis, A. (2002b), "Organic product avoidance: reasons for rejection and potential buyers' identification in a countrywide survey", British Food Journal, Vol. 104, 233-260.

Green, J.M., Draper, A., Dowler E.A., Fele, G., Hagenhoff, V., Rusanen, M., and Rusanen, T. (2005), "Public understanding of food risks in four European countries: a qualitative study", European Journal of Public Health, Vol. 15 (5), 523-527.

Gurviez, P. (1999), "La confiance comme variable explicative du comportement du consommateur: proposition et validation empirique d'un modèle de la relation à la marque intégrant la confiance », Proceedings of the FMA International Congres, Strasbourg, France.

Gurviez, P., and Korchia, M. (2002), "Proposition d'une échelle de mesure multidimensionnelle de la confiance dans la marque", Recherche et Applications en Marketing, Vol. 17 (3), 41-61.

Hair, Joseph F., William C. Black, Barry Babin, J., Rolph E. Anderson, and Ronald L Tatham (2006), Multivariate Data Analysis (6th ed.). Upper Saddle River, N.J.: Pearson Education Inc.

Hamzaoui L. and Zahaf M. (2008), "Profiling Organic Food Consumers: Motivations, Trust Orientations and Purchasing Behavior", Journal of International Business and Economics, Vol. 8 (2), 25-39.

Hamzaoui L. and Zahaf M. (2009), "Exploring the decision making process of Canadian organic food consumers: motivations and trust issues", Qualitative Market Research, Vol. 12 (4), 443-459.

Hughner, R.S., McDonagh, P., Prothero, A., Shultz II, C.J., and Stanton, J. (2007), "Who are organic food consumers? A compilation and review of why people purchase organic food", Journal of Consumer Behaviour, Vol. 6, 94-110.

Hartman Group and New Hope. 1997. *The Evolving Organic Marketplace*. Bellevue, WA.

Jones, P., Clarke-Hill, C., Shears, P. and Hillier, D. (2001), "Case study: retailing organic foods", British Food Journal, Vol. 103 (5), 359-65.

Kassarjian H. (1977). Content Analysis in Consumer Research, *Journal of Consumer Research*, Vol. 4 (1), p. 8.

Kristallis, A., and Chryssohoidis, G. (2005), "Consumers' willingness to pay for organic food: Factors that affect it and variation per organic product type", British Food Journal, Vol. 107 (5), 320-343.

Laroche, M., Bergeron, J., and Barbaro-Forleo, G. (2001), "Targeting consumers who are willing to pay more for environmentally friendly products", Journal of Consumer Marketing, Vol. 18 (6), 503-520.

Larue, B., West, G., Gendron, C., and Lambert, R. (2004), "Consumer response to functional foods produced by conventional, organic, or genetic manipulation", Agribusiness, Vol. 20 (2), 155-166.

Macey, A. (2003), "Certified Organic, The Status of the Canadian Organic Market in 2003", Report to Agriculture & Agri-Food Canada.

Macey, Anne (2007). "Retail Sales of Certified Organic Food Products in Canada in 2006" (pdf). *Organic Agriculture Center of Canada*. http://www.organicagcentre.ca/Docs/RetailSalesOrganic_Canada2006.

Millock, K., Hansen, L.G., Wier, M., and Andersen, L.M. (2002). Willingness to pay for organic foods: a comparison between survey data and panel data from Denmark. (online). Available at: http://www.akf.dk/organicfoods/conference/willingness.pdf. Proceedings of the Conference on Consumer Demand for Organic Foods, Denmark.

Mitchell, V.W., and McGolrick P.J. (1996), "Consumer's risk-reduction strategies: a review and synthesis", The International Review of retail, Distribution and Consumer Research, Vol. 6 (1), 1-33.

Organic Monitor (2006), The European Market for Organic Food and Drink, Organic Monitor, London.

Padel, S., and Foster, C. (2005), "Exploring the gap between attitudes and behaviour: Understanding why consumers buy or do not buy organic food", British Food Journal, Vol. 107 (8), 606-625.

Rao, A. R. and Bergen, M.E. (1992), "Price premium variations as a consequence of buyers' lack of information", Journal of Consumer Research, Vol. 15, 235-264.

Rostoks, L. (2002), "Romancing the organic crowd: this new category may yield plenty of profits for you, if you master the new merchandising rules to attract the organic consumer", Canadian Grocer, Vol. 116, 22-24.

Saha, M & Darnton, G. (2005), "Green companies or green con-panies: Are Companies Really Green, or Are They Pretending to Be?", Business & Society Review, Vol. 102 (2), 117-157.

Sheperd, R., Magnusson, M. and Sjoden, P.O. (2005), "Determinants of consumer behaviour related to organic foods", Ambio, Vol. 34 (4/5), 352-359.

Sassatelli, R. and Scott, A. (2001), "Novel Food, New Markets and Trust Regimes - Responses to the erosion of consumers' confidence in Austria, Italy and the UK", European Societies, Vol. 3 (2), 213-244.

Sirieix, L., Pontier, S., and Schaer, B. (2004), "Orientations de la confiance et choix du circuit de distribution: le cas des produits biologiques", Proceedings of the 10th FMA International Congres, St. Malo, France.

Sirieix, L., Pernin, SJ.-L., and Schaer, B. (2009), "L'enjeu de la provenance régionale pour l'agriculture biologique ", Innovations Agronomiques, Vol. 4, 401-407.

Smithers, J., Lamarche, J., and Alun, J. (2008), "Unpacking the terms of engagement with local food at the Farmers' Market: Insights from Ontario", Journal of Rural Studies, Vol. 24 (3), 337-350.

Thilmany, D., Umberger, W. and Ziehl, A. (2006), Strategic market planning for value-added natural beef products: A cluster analysis of Colorado consumers. Renewable Agriculture and Food Systems (2006) 21: 192-203 accessed on-line http://www.organicagcentre.ca/ResearchDatabase/res_marketing_beef.asp

Torjusen, H., Sandstad, L., O'Doherty Jensen, K., and Kjaernes, U., (2004), "European consumers' conceptions of organic food: A review of available research", Professional report, (4).

Tutunjian, J. (2004), "Are organic products going mainstream?", Canadian Grocer, Vol. 118, 31-34.

Tutunjian, J. (2008), "Market survey 2007", Canadian Grocer, Vol. 122 (1), 26-34.

Verdurme, A., Gellynck, X., & Viaene, J. (2002), "Are organic food consumers opposed To GM food consumers?", British Food Journal, Vol. 104 (8), 610-623.

Van Elzakker B. and Eyhorn F. (2004), "Developing sustainable value chains with smallholders", The Organic Business Guide, Online Report.

Vlosky, R.O., Ozanna, L.K., and Fontenot, R.J. (1999), "A conceptual model of US consumer willingness-to-pay for environmentally certified wood products", Journal of Consumer Marketing, Vol. 16 (2), 122-136.

Wier, M., and Calverly, C. (2002), "Market potential for organic foods in Europe", British Food Journal, Vol. 104 (1), 45-62.

Willer, H., and Yussefi (2007), "The World Of Organic Agriculture", Online Report.

Willer, H., and Kilcher, L. (2011), "The World of Organic Agriculture. Statistics and Emerging Trends 2011", IFOAM, Bonn, & FiBL.

Worner, F. and Meier-Ploeger, A. (1999), "What the consumer says", Ecology and Farming, Vol. 20, January-April, 14-15.

Yiridoe, E.K., Bonti-Ankomah, S., & Martin, R.C. (2005), "Comparison of consumer perceptions and preference toward organic versus conventionally produced foods: A review and update of the literature", Renewable Agriculture and Food Systems, Vol. 20 (4), 193-205.

Zanoli, R., and Naspetti, S. (2002), "Consumer motivations in the purchase of organic food: a means end approach", British Food Journal, Vol. 104 (8), 643-653.

Zander, K., and Hamm, U. (2010), "Consumer preferences for additional ethical attributes of organic food", Food Quality and Preferences, 21, 495-503.

Zepeda, L., & Li, J. (2007), "Characteristics of Organic Food Shoppers", Journal of Agriculture and Applied Economics, Vol. 39 (1),17-28.

Zepeda, L., and Deal, D. (2008), "Think before you eat: Photographic food diaries as intervention tools to change dietary decision-making and attitudes." *International Journal of Consumer Studies*. (32), 692-698.

5

University Student Attitudes Toward Organic Foods

Aslı Uçar and Ayşe Özfer Özçelik
Ankara University/Faculty of Health Sciences Department of Nutrition and Dietetics
Turkey

1. Introduction

In the recent years, new food trends have emerged, among the most popular ones of which is the organic trend. Many food companies have begun offering organic products similar to many conventional products that have been stocked on grocery shelves. These organic foods have gained a share of two percent in the grocery store purchases. Soon after organic foods began to appear, natural foods have come into the spotlight. The labels of these products read "no preservatives", "no artificial colors" or "no artificial flavors" (Solano, 2008).

Organic agriculture is not the same as natural agriculture; the former does not use fertilizers or agricultural pesticides. Organic agriculture is an ecological production method that increases and develops the value of biological activities and varieties. Organic foods are subject to a process during which no synthetic fertilizers, agricultural pesticides, chemical herbicides (that control harmful weeds), hormones, pigments, antibiotics or chemical metal packaging etc. are used (Shepherd et al., 2005).

Organic farming is carried out in at least 160 countries worldwide. The share of organic agricultural land in total agricultural land as well as the number of organic holdings is continuously growing. The organic products market is also growing not only in Europe and North America, where the largest markets are located, but also in many emerging economies and economies in transition. More than thirty-seven million hectares of agricultural land are managed organically by 1.8 million producers. About one-third of the world's organic land -13.4 million hectares- is located in emerging markets and markets in transition (SIPPO, 2011). Top ten countries which give priority to the organic food production in the total agricultural land are Liechtenstein (26.9%), Austria (18.5%), Sweden (12.6%), Switzerland (10.8%), Estonia (10.5%), Czech Republic (9.4%), Latvia (9.0%), Italia (8.7%), Slovakia (7.5%) and Finland (6.6%) (Willer, 2011). In 2009 global sales of organic food products reached 55 billion US dollars, more than doubling in value from 25 billion US dollars in 2003. Europe was the second largest organic market in the world after the USA with a turnover of 26 billion US dollars in 2009. Germany had 5.8 billion euros, followed by France with 3 billion euros, the UK (2.1 billion), Italy (1.5 billion) and Switzerland (1 billion). These figures increase year by year and impressively illustrate the powerful development of the organic food production and market all over the world (SIPPO, 2011).

Organic farming in Turkey, as distinct from the developments in Europe, commenced as export oriented activities in 1986 in line with the requests of importing companies. At first production and exports were conducted suitable to the legislation of importing countries relating to this subject but continued in line with the Regulations of the European Community after 1991. In the Commission Regulation (EEC) 94/92 dated 14 January 1992 as a supplement to Regulation No. 2092/91, detailed information was provided on compliance by countries that export organic products to the European Community, that these countries put into practice the legislation of these nations and a file containing various technical and administrative matters that were included within this legislation were made compulsory in export applications to the European Community.

In order to harmonize with these developments in the European Community, the Ministry of Food, Agriculture and Livestock, together with the cooperation of various organizations and corporations commenced with preparatory activities for a regulation and the "Regulation Relating to the Production of Agricultural and Livestock Products by Ecological Methods" came into force with its publication in the Official Gazette No. 22145 of 24.12.1994. Revisions were made to this Regulation in 1995 and subsequently a new Regulation was published in 2002.

The draft law relating to the production, consumption and inspection of organic products took place within the emergency action plan of the Government and "No. 5262 on Organic Farming Law" was published in the Official Gazette No. 25659 dated 03.12.2004. The "Regulation Relating to the Procedures and Implementation of Organic Farming" prepared in accordance with Law No. 5262 came into force with its publication in the Official Gazette No. 25841 of 10.06.2005. This regulation was revised three times bearing in mind harmonization with EU legislation and the conditions pertaining to each country.

The "Regulation Relating to the Procedures and Implementation of Organic Farming" that is harmonized with the new legislation of the EU came into effect with its publication in the Official Gazette No. 27676 dated 18.08.2010 (Anonymous, 2011a).

Ecological farming is a production activity that is subject to controlled and certification. Procedures involving control and certification on ecological farming activities in Turkey are conducted by private corporations authorized by the Ministry of Food, Agriculture and Livestock in accordance with Article 11 of the Organic Farming Law.

Based on this Article, domestic and/or foreign private and official organizations that wish to carry out control and certification procedures in Turkey must apply to the Ecological Farming Commission of the Ministry of Food, Agriculture and Livestock. Prior to this application, foreign organizations must obtain the appropriate work permit from the Prime Ministry Undersecretariat of the Treasury. The Ecological Farming Commission issues an ecological farming supervision and certification permit for a particular time period after reviewing and approving the organization in question.

Since organic farming is in the form of contractual farming, control is regular record keeping of production from start to finish, observing the production process, reporting the results of the observations and testing and supervising the organic qualities of the product through laboratory analysis.

The product obtained using organic farming methods by the supervised grower, determining whether it is of organic quality and registration of certification are undertaken by authorized supervision and certification organization. These organizations receive their authority from the Ecological Farming Commission of the Ministry of Food, Agriculture and Livestock.

Up to the present time, the authority to control and certify organic farming in Turkey on behalf of the Ministry has been given to 17 control organizations. While on the one hand these organizations provide for organic production suitable to certain criteria, controlled at every stage and for these products to be of quality and certified, they are, on the other hand, in a position to conduct the necessary chemical, microbiological and all other types of analysis from samples and specimens taken from the place of production and after evaluating the results, to label them according to their suitability. All of these activities are carried out on behalf of the Ministry and thus, these organizations are accountable to the Ministry itself (Kirazlar, 2001). At present an accredited laboratory that can carry out these analyses is only present in Izmir. This laboratory is currently conducting activities to increase the number of active substances analyzed as well as accredited active substances (Anonymous, 2011b).

In Turkey, according to the overall organic agricultural production data, the number of organic foods produced is 150 in 2002 while this number increased to 216 in 2010. In the recent years there has been a clear increase in the number of workers dealing with organic agriculture (from 12.428 people in 2002 to 42.097 people in 2010) as well as in the production space (from 89.827 hectares in 2002 to 510.033 hectares in 2010) (Anonymous, 2011c). The export rates in organic products show varieties throughout the years; in 2002 the export revenue was $ 30.877.140 while in 2010 it receded to $ 15.879.571 and the amount of export products decreased (Anonymous, 2011d) in spite of the increase in overall production. This can be interpreted as an increase in domestic consumption.

Although the internal market concerning organic products is not well established in Turkey, these products are sold in stands in big shopping malls. Furthermore, especially in big cities there are organic-only stores. The customer profile of these stores consists of persons with higher education, higher income, of middle and upper middle ages (Kaya, 2003). The increase in food and nutrition related health problems are directing consumers towards organic products. Due to the disparities in existing income levels in Turkey, the structure of consumption demand is adversely affected. According to estimates, the number of retail greengrocers that sell organic products in Turkey is close to one hundred. In addition to these greengrocers, certain supermarkets have organic product aisles. With consumer awareness and an increase in demand, these numbers are expected to increase. Certain hotels, motels and restaurants in touristic locations are showing signs of an increase in organic foods on their menus from natural and organic products. At present the organic domestic market in Turkey is very small but the production capacity for organic products is much higher than domestic market demand and the Turkish food industry has the adequate means, knowledge and experience on this subject (Atay and Sarı, 2007).

The researches conducted in different countries show that organic food consumers in general are educated women ranging between 30-49 years of age, with high incomes and children of six years of age or older (Davies et al., 1995; Essoussi & Zahaf, 2008; Tsakiridou et al., 2008). Makatouni (2002) states that values related to the health of one's self or family, the environment and the animal welfare are responsible for choosing organic foods. Baker et al. (2004) also claims that both German and British organic food consumers are concerned with "health, well-being and the enjoyment of life", but that the food attributes "achieving" this concern are different. For British consumers, "healthiness" and "non-genetically modified" attributes are important, whereas Germans seek "taste" and "quality". Schifferstein & Oude Ophuis (1998) find out that health was an important motivator for buying organic foods for "incidental" organic buyers in the Netherlands, whereas both health and environmental concerns were regarded by "regular" buyers. Perceptions about

organic foods remain the same; that they are healthy, beneficial for the environment and taste good or better than conventional foods (Lea & Worsley, 2005; Padel & Foster, 2005; Zhao et al., 2007).
Consumers interested in the organic food industry are also worried about pesticides and growth hormones in food. The organic food sector's main target is individuals aging from 18 to 29 years; however, this group of people does not consist of regular buyers. Tsakiridou et al. (2008) finds out that most college students do care about the environment and are interested in organic foods, but at this point in their lives they cannot justify spending higher prices on organic foods; while most organic food consumers have college degrees and are interested in the processing and handling of the food they consume. Younger generations are the original target but research has shown that women in their 30s to 50s form the majority of individuals buying organic products (Essoussi & Zahaf, 2008).
Many conventional products today have their organic equivalents. Lately natural foods have begun to appear on grocery shelves alongside conventional and organic products. However, it may not be clear to the average consumer what differences there are between these products. If consumers do not understand how these products differ they may be purchasing a product with attributes that they do not want to pay for. They may be paying a higher price than necessary (Solano, 2008). This study has been planned focusing on the assumption that education has certain effects on the individuals' behavior and attitude, and it was carried out among the university students in Ankara in order to determine their attitudes towards organic foods.

2. Methodology

2.1 Procedure

The center of the research was Ankara, Turkey's capital, and the research participants were the students attending Ankara University Faculty of Health Sciences. The reasons behind choosing this faculty as the site of research are that the researchers is working at the same faculty, and that the students attending may have more information on organic foods and answer the statements in the questionnaire more consciously thanks to the curriculum of their faculty. The faculty consists of six departments, namely the Department of Nutrition and Dietetics, Child Development, Midwifery, Nursery, Health Services Management and Social Work. The Department of Child Development is excluded from the research as there are only freshmen students attending the department. All of the departments in the faculty are four-year schools. In the Faculty of Health Sciences there are 399 freshmen students, 331 sophomores, 302 juniors and 216 seniors (Department of Child Development excluded), the total of which amounts to 1248 students. This number consists of 1025 female students to 223 male students. In the beginning of this research it was planned to include all of the students attending to the faculty. The questionnaires were handed out to everyone who had come to school that day and delivered to 879 people. 40 questionnaire forms are excluded from the research due to being filled incorrectly or incompletely. So, 849 students -145 of which are from Department of Nutrition and Dietetics, 226 from Department of Midwifery, 207 from Department of Nursery, 140 from Department of Social Services, 121 from Department of Health Services Management- have participated in the research.
The research data is collected through a questionnaire along with face to face interviews by researchers. Previous studies on the topic are utilized in preparing the questionnaire items

(Wen Chei Chan, 2008; Vanderkloet, 2008; Chen, 2007; Lawrence, 2007; Lockie et al., 2004). The questionnaire form is composed of two sections, the first of which contains general information items; the second part contains items related to the attitudes towards organic foods. The data was collected between April and May 2011.

A Likert type scale consisting of 59 statements was used to determine the perception of organic foods. There were 59 statements in the scale, including both negative and positive sets of sentences. The positive statements were scored as follows: "I strongly agree" (5), "I agree" (4), "Undecided" (3), "I don't agree" (2), and "I strongly disagree" (1). The scoring for the negative statements was exactly the opposite (statements no: 9, 17, 18, 23, 24, 26, 40, 44, 45, 47, 51, 52).

Whether the items in the questionnaire were able to measure the attitudes towards organic foods or not was tested by using structural validity analysis. The analysis determined the items that measured the repeating and different structures, and whether the items were included in a sub-structure or not was determined by examining the values for item factor loadings (Büyüköztürk, 2002).

Although loading values of 0.45 and above are recommended in the factor analysis, in practice, there has been some cases in which a loading value of 0.30 was also acceptable as the lowest loading value. In this study, too, the loading value of 0.30 or above for an item was accepted as adequate. The items with higher values were selected, and those with lower values were not included in the statistical analyses applied in the later stages of the study (18 items were removed) (Kerlinger, 1973; Tabanchinck & Fidell, 1989).

For the reliability of the questionnaire, "Cronbach Alpha", the internal consistency coefficient is calculated, and the alpha value was found to be 0.93 (mean: 3.74, min: 3.13, max: 4.21). Accordingly, it is agreed that the "Attitudes towards Organic Foods" scale was a valid and reliable instrument.

The factor analysis conducted for the validity of the "attitudes towards organic foods" scale results in six factors: (i) Positive statements regarding organic foods (factor loadings 0.37-0.74); Negative statements regarding organic foods (factor loadings 0.53- 0.70); Statements regarding organic food production (factor loadings 0.60-0.70), Purchasing organic foods (factor loadings 0.54-0.68), Comparing organic foods with traditional foods (factor loadings 0.37-0.67), Organic foods and environment/chemical usage (factor loadings 0.49-0.73) (see Appendix).

When all the statements were replied, the grades obtained from the 'Positive statements regarding organic foods' part amounts to 60 points; the 'Negative statements regarding organic foods' and 'Comparing organic foods with traditional foods' parts, 40 points; the 'Statements regarding organic food production' and 'Purchasing organic foods' parts, 20 points, and the 'Organic foods and environment/chemical usage' part, 25 points, sum total of which is 205 points.

2.2 Statistical analyses

Data obtained as the result of the research were evaluated by the Statistical Package for Social Sciences (SPSS) software, by taking the "gender", "department", "grade" variables into consideration. In evaluating the organic food attitudes grades, "Independent- samples *T* test" for the gender variable, "One-way anova" analysis and "LSD test" for the other variables have been applied. Frequencies, averages and standard deviations were calculated (Kesici & Kocabaş, 1998).

3. General Information

3.1 Demographics

Seventeen point five percent of the students participating in the research are males while 82.5% are females. Seventeen point three percent of the students are from the Department of Nutrition and Dietetics, 26.9% from Midwifery, 24.7% from Nursery, 16.7% from Social Services, 14.4% from Health Services Management. Thirty two point eight percent of the students are freshmen, 32.9% sophomores, 24.7% juniors and 9.7% seniors. Forty nine point six percent of the students stay in the dormitories while 30.3% share home with their friends, 17.6% stay with their families and 2.5% live alone or with relatives.

3.2 Organic food consumption

When asked whether they make shopping for the household, 51.6% of the students answered "sometimes", 46.1% "always" and 2.3% "never". Fifty seven point four percent of the students stated that they eat organic food. Those who do not consume organic food gave the reasons that they cannot find the products everywhere (45.7%), the products are expensive (38.7%), they do not have the need for these products (15.7%). Fifty six percent of the students which claimed to eat organic food consumes these products occasionally, while 25.1% of them answered "frequently", 16.0% "rarely" and 2.9% "always". They completed the statement "If I could not find organic food at the store I go for shopping..." and 42.7% of the answers are "I would buy non-organic foods", 25.9% are "I would buy another similar organic food", 15.6% are both "I would go to another store that I possibly could find" and "I would postpone shopping". The statement "If the prices between organic foods and traditional foods are similar...", asked to determine the effect of the prices on students' preferences, is mostly (52.3%) answered with "I would buy as much organic food as I can". Twenty eight point four percent of the answers are "I would do whatever I am doing currently" while 19.3% are "I would buy more organic food than I do currently".

3.3 Attitudes towards organic foods

In Table 1 the students' answers to the statements regarding organic foods are shown in percent values. Although most of the students think that organic foods have better quality and are free of any unhealthy effects, they have shown ambivalence (46, 48, 55) or disagreement (29, 36, 38, 42, 49, 59) to certain statements. Among the positive statements presented, the one which has been disagreed most (55.3%) is "I would buy organic foods as I see many benefits in them (for environment, health and animal rights)". Those who strongly agree with the statement "I feel healthy when I eat organic foods" are the fewest (2.2%).

Nevertheless, the students show disagreement to all of the negative statements regarding organic foods. Most of the students participating in the research have correct knowledge about the production of the foods (that they do not contain additives, preservatives/artificial flavors, are not treated with hormones or antibiotics, and are not genetically modified). They mostly answer "I agree" to the statements in this section.

Although they find organic foods more expensive, the rate of those who think that more agricultural land and more space in the groceries should be given to organic foods and of those who would be glad to consume more organic food is higher. According to the statements given in the section in which organic foods are compared with traditional foods, the students find organic foods more reliable in terms of taste, nutrition, health, quality, safety and against the risk of illnesses. Except for the statement "Organic foods reduce soil contamination", the students mostly answered "I agree" to the statements related to organic foods and chemical usage.

Factor 1 Positive statements regarding organic foods	I strongly agree	I agree	Undecided	I don't agree	I strongly disagree
20. Organic foods have better quality. (n=835)	26.0	47.2	20.7	4.0	2.2
22. Organic foods are free of harmful effects. (n=829)	14.2	42.7	32.1	8.2	2.8
29. I feel healthy when I eat organic foods. (n=830)	2.2	10.0	26.4	40.8	20.6
36. I think organic foods are fresher. (n=838)	2.3	7.9	21.8	46.4	21.6
38. I would buy organic foods as I see many benefits in them (for environment, health and animal rights). (n=835)	1.4	7.8	17.8	55.3	17.6
42. If everyone consumed organic products, the world would be a better place. (n=831)	2.8	10.6	27.1	40.4	19.1
46. Organically produced food means completely reliable food. (n=824)	4.9	19.7	40.4	25.0	10.1
48. Infant foods with organic ingredients have more nutrients than traditional foods. (n=832)	4.1	14.2	38.1	31.2	12.4
49. I live longer if I consume organic foods. (n=831)	5.8	17.2	31.5	33.0	12.5
50. Consuming organic foods helps the prevention of obesity. (n=830)	2.8	12.8	28.6	40.4	15.5
55. I think organic food consumption has no risks. (n=830)	3.0	15.7	39.3	28.8	13.3
Factor 2 Negative statements regarding organic foods					
24. I am against my parents' or relatives' consumption of organic foods. (n=821)	3.0	4.9	7.9	46.2	38.0
26. I do not consume organic foods because I do not want to try foods produced by new technologies. (n=821)	4.0	8.9	19.0	47.1	21.0
40. Consuming organic foods are not amongst the actions I can take to protect the environment. (n=837)	5.0	14.7	24.4	43.5	12.4
44. Organic infant foods are not as healthy as traditional foods. (n=828)	5.9	12.7	35.1	34.1	12.2
45. Organic foods will bring harm to the society rather than benefits. (n=826)	3.4	6.5	18.5	47.8	23.7
47. As it modifies the foods we consume, it is dangerous to eat organic foods. (n=833)	3.7	11.3	26.2	42.1	16.7
51. I think the risks organic foods bring surpass their benefits. (n=827)	5.7	11.9	27.7	41.1	13.7
52. Regular consumption of organic foods is harmful for my health. (n=825)	3.0	7.2	22.3	48.1	19.4

Factor 3 Statements regarding organic food production					
14. Organic foods do not contain additives.	24.2	41.5	23.7	8.5	2.1
32. Organic foods do not contain preservatives or artificial colors. (n=823)	20.8	42.0	27.6	7.4	2.2
33. Organic foods are produced without using hormones or antibiotics. (n=830)	17.8	37.6	33.7	8.1	2.8
34. Organic foods are not genetically modified. (n=836)	21.9	39.6	28.6	7.3	2.6
Factor 4 Purchasing organic foods					
13. I would be glad to consume more organic food as long as I can find them. (n=826)	40.8	41.3	10.8	4.6	2.5
15. I think groceries should give more space to organic foods. (n=833)	40.2	44.1	9.1	4.7	1.9
16. I think more agricultural land should be spared for organic foods. (n=829)	42.3	43.4	7.5	3.7	3.0
21. Organic foods are more expensive. (n=830)	37.3	46.9	10.8	3.5	1.4
31. I am protecting the environment by consuming organic foods. (n=834)	17.3	47.7	26.4	7.0	1.7
Factor 5 Comparing organic foods with traditional foods					
6. Organic foods taste better than traditional foods. (n=825)	38.8	35.9	15.3	7.5	2.5
7. Organic foods are more nutritious than traditional foods. (n=824)	37.7	36.5	16.1	6.8	2.8
8. Organic foods reduce the risk of illnesses. (n=809)	40.8	40.5	12.0	4.7	2.0
9. The vitamin and mineral content of organic foods are not more than those of traditional foods. (n=824)	7.6	16.4	22.5	37.5	16.0
10. Organic foods are more reliable than traditional foods. (n=825)	32.4	40.2	19.5	5.8	2.1
17. I do not think there is any difference between organic foods and traditional foods. (n=830)	4.3	8.8	17.8	44.7	24.3
18. Organic foods do not have better quality than traditional foods. (n=821)	2.6	9.4	19.4	47.1	21.6
23. I think the consumption of organic foods is a trend. (n=828)	11.0	22.9	19.9	32.5	13.6
Factor 6 Organic foods and chemical usage					
1. Organic foods are not treated with any chemicals. (n=832)	22.2	45.0	19.5	10.5	2.9
2. No chemical fertilizer is used in the production of organic foods. (n=834)	19.2	37.3	26.9	12.6	4.1
3. Organic foods reduce soil contamination. (n=835)	2.2	4.0	20.7	47.2	26.0
4. Organic foods reduce the use of pesticides in agriculture. (n=800)	15.6	38.2	37.9	6.1	2.1
5. Organic foods secure the biological balance of nature. (n=821)	29.2	53.0	11.3	4.5	1.9

Table 1. Students' attitudes towards organic foods (%)

3.4 Variables

The findings have been analyzed with respect to gender, department and grade variables.

3.4.1 Gender

In Table 2 average points related to the students' attitudes towards organic foods are presented.

	Male	Female	t	Sig
Positive statements regarding organic foods	41.0±8.20	42.6±7.21	2.300	0.022*
Negative statements regarding organic foods	27.2±5.83	29.0±5.34	3.460	0.001**
Statements regarding organic food production	14.0±3.65	14.7±3.20	2.223	0.027*
Purchasing organic foods	15.3±3.69	16.7±2.78	4.309	0.000***
Comparing organic foods with traditional foods	28.4±6.10	18.9±5.46	2.689	0.007**
Organic foods and chemical usage	17.1±4.33	29.8±3.45	4.307	0.000***
General attitudes	143.3±23.6	151.6±20.9	3.977	0.000***

*p<0.05, **p<0.01, ***p<0.001

Table 2. The results of t test towards organic food attitudes scale of students based on gender

In the general total, the average points of females (151.6±20.9) are higher than males (143.3±23.6). According to the statistical analysis results this difference is important (p<0.001). When the chart is evaluated regarding six different parts, except for the part "Comparing organic foods with traditional foods", in each part the same pattern can be found although the magnitude scale shows differences.

3.4.2 Department

Table 3 shows the average points of the students' attitudes towards organic foods for each department

	1	2	3	4	5	F	Sig	Difference
Positive statements regarding organic foods	41.4±6.24	43.5±8.07	43.0±6.49	40.9±8.05	41.6±7.79	4.025	0.003*	1-2, 2-4, 2-5, 3-4
Negative statements regarding organic foods	30.4±4.20	28.3±5.85	29.2±5.21	28.4±5.78	27.2±5.62	6.630	0.000**	1-2, 1-3, 1-4, 1-5, 3-5
Statements regarding organic food production	14.7±2.88	14.9±3.46	14.6±3.22	14.1±3.26	14.1±3.54	1.717	0.144	
Purchasing organic foods	16.9±2.33	16.7±3.06	16.7±2.74	15.8±3.22	15.5±3.51	5.994	0.000**	1-4, 1-5, 2-4, 2-5, 3-4, 3-5

Comparing organic foods with traditional foods	29.9±5.02	30.0±5.66	30.2±5.36	29.2±5.48	27.6±6.27	5.221	0.000**	1-5, 2-5, 3-5, 4-5
Organic foods and chemical usage	19.2±3.14	18.9±3.56	18.7±3.50	17.3±4.00	17.7±3.96	7.606	0.000**	1-4, 1-5, 2-4, 2-5, 3-4, 3-5,
General attitudes	152.8±18.5	152.5±23.0	152.5±19.4	146.0±22.6	143.4±22.8	6.212	0.000**	1-4, 1-5, 2-4, 2-5, 3-4, 3-5

1. Nutrition and Dietetics 2. Midwifery 3. Nursery 4. Social Services 5. Executive in Health Institutes
*p<0.01, **p<0.001

Table 3. The results of variance analyses of the attitude scales towards organic foods based on students' departments

The department with the lowest score in the attitude scale towards organic foods is Health Services Management (143.4±22.8). It is followed by the points scored by the students in the Department of Social Work (146.0±22.6). According to the results of the statistical analysis results this situation is meaningful (p<0.001) and the difference is generally between these two departments and the rest. Other three departments have scored quite similarly, the highest of which belongs to the students in the Department of Nutrition and Dietetics.

3.4.3 Grade

Table 4 shows the average attitude points towards organic foods based on the students' grade

	1	2	3	4		Sig	Difference
Positive statements regarding organic foods	38.9±6.97	38.5±6.75	37.7±6.53	40.2±7.79	2.963	0.031*	2-4, 3-4
Negative statements regarding organic foods	28.8±5.52	28.7±5.48	28.1±5.23	30.2±5.60	3.048	0.028*	1-4, 2-4, 3-4
Statements regarding organic food production	14.6±3.38	14.6±3.14	14.1±3.23	15.0±3.62	1.711	0.163	
Purchasing organic foods	19.9±3.56	20.5±3.19	19.6±3.65	20.8±3.59	4.052	0.007**	1-2, 1-4, 2-3, 3-4
Comparing organic foods with traditional foods	29.8±5.51	29.5±5.37	29.0±5.73	30.2±6.30	1.213	0.304	
Organic foods and chemical usage	18.7±3.45	18.2±3.58	18.4±3.78	18.7±4.35	0.866	0.458	
General attitudes	150.7±22.7	150.3±19.9	147.2±20.5	155.4±24.8	2.990	0.030*	3-4

*p<0.05, **p<0.01

Table 4. The results of variance analyses towards organic food attitudes scale of students based on grade

In the general total, third-grade students scored the lowest, and the difference of their score with senior students' score is considerable ($p<0.05$). When all departments evaluated, the attitude scales of the fourth-grade students are always higher. The scores attained from the parts positive statements regarding organic foods ($p<0.05$), negative statements regarding organic foods ($p<0.05$) and purchasing organic foods ($p<0.01$) show statistically considerable difference by grades. In general this difference can be considered as resulting from the scores attained by fourth-grade students.

4. Discussion

This study aims to determine future-consumer students' attitudes towards organic foods the production and consumption of which increase day by day. After the evaluations it is concluded that the students' attitudes are in general positive. In various studies, too, similar results have been obtained (Chrysshoidis &Krystallis, 2005; Tarkiainen & Sundqvist, 2005; Lawrence, 2007; Arvola et al., 2008; Urena et al., 2008).

Several factors affect the development of positive attitude towards any food. Various studies have shown that concerns about health and environment also bring about positive attitudes towards organic foods (Chrysshoidis & Krystallis, 2005; Radman, 2005; Saba & Messina, 2003; Magnusson et al., 2003; et al., 1994; Davies et al. 1995; Wandel & Bugge, 1997). Chen (2007) states, in the study conducted in Taiwan, that the naturalness of the food, animal rights and protection of the environment are also among the sources of positive attitudes towards organic foods for the adults.

Regular consumers of organic products believe that organic food is healthier and has a better quality than factory or traditionally processed food (Vindigni et al., 2002). Bissonnette and Contento (2001) find out in their study that about three-quarters of the respondents agreed that organic foods are better for the environment (73.7%) and better for their personal health (74.8%). Radman (2005) deduces that Croatian consumers consider organically grown products as very healthy and of good quality. Another study conducted in Canada shows that the participants find organic foods healthier and more beneficial for the environment (The Nielsen Company, 2007).

Lawrence (2007) finds out that college students at Oklahoma State University agrees with many statements presenting organic foods as a healthy, non-risky, environmentally friendly food option. These students perceive organic foods to be completely safe to eat and are willing to serve organic foods to their friends.

The attitudes and thoughts of people towards organic foods have effects on buying decisions. According to a survey, 52.8% of Americans buy organic foods because they believe organic foods are better for their health and 52.4% buy them because they believe these foods are better for the environment (Whole Foods Market, 2005). Hughner et al. (2007) shows that consumers' main reason for purchasing organic food is health concerns. In a study conducted in Northern Ireland, it is found that the three main reasons for purchasing organic foods are health, environment and taste respectively (Davies et al., 1995).

Consumers believe that organic products have better quality; therefore, these products also will taste better (Roddy et al., 1996). Taste is a factor affecting customers' choice in purchasing organic products, but studies have shown that they detect no difference in taste when comparing organic meats to non-organic meats (McEachern & Willock, 2004). But Radman (2005) finds out that Croatian consumers consider organically grown products very tasty. In our study, too, the participating students find these foods more delicious and healthy and think that their taste is better (Table 1).

Generally if one has a positive attitude towards something, this attitude is expected to reflect in the behavior later (Ajzen, 1991). In other words, if a person has a positive attitude towards organic foods, he/she is expected to buy them. However, Shepherd et al. (2005) found a disparity between attitudes and behavior - despite the majority of surveyed consumers holding positive attitudes towards organic foods, only 4-10% reported an inclination to choose the organic options next time. For organic foods are also thought to be expensive (Finch, 2005; Padel & Foster, 2005; Radman, 2005; Shepherd et al., 2005; Whole Foods Market, 2005; Anonymous, 2004) - a common barrier to their purchase (Finch, 2005; Fotopoulos & Krystallis, 2002; Lea & Worsley, 2005; Padel & Foster, 2005). In another study conducted in Turkey, the reasonability of the prices ranked highest among the attributes sought in organic products (Sarıkaya, 2007). There are also discouragements to buying organic food, mainly high price and low availability (Fotopoulus & Krystallis, 2002; Anonymous, 2004; Lea & Worsley, 2005; Padel & Foster, 2005; Tsakiridou et al., 2008). In our study, it is shown that the students find organic foods more expensive and they state that they would buy more organic food if they had a similar price with traditional food. Considering the per capita income in Turkey, these statements are reasonable. Resembling to our results, in the study conducted by Chrysshoidis and Krystallis (2005), the participants state that they like organic foods better than traditional foods and they would prefer them if sold in similar prices.

There are several factors involved in the choice of organic foods: place of purchase, values, beliefs that they are healthier or better for the environment, positive attitudes, and sensory characteristics of the food. In addition to these factors, certain demographic characteristics, such as being female or young, may also contribute to a greater likelihood of buying organic food (Davies et al., 1995; Lockie et al., 2004; Onyango et al., 2007). Consumers of organic products are typically women in their late 30s to 50s, and normally these consumers are interested in their health and uncomfortable about food not organically raised; they generally have children and at least a bachelor's degree (Davies et al., 1995). These consumers are concerned with animal welfare, environmental issues and knowing how their food is raised (Bellows et al., 2008). Radman (2005), Storstad & Bjørkhaug (2003) and Davies et al. (1995) have shown in their studies that females buy more organic foods than males. In this study, too, female students give more positive answers to all statements about their attitudes towards organic foods than male student.

Bocaletti (2009) states that students, housewives, highly-educated people and those concerned with environment have positive attitudes towards organic foods. Some studies show that the young have positive attitudes towards organic foods, but the actual consumers consist of individuals of older age groups (Hughner et al., 2007; Magnusson et al., 2001). In our study, too, more than half of the participants claim to consume organic foods while only 2.9% of them are regular consumers. This low rate may be the result of certain circumstances such as the study group's consisting of students, their financial dependence on their families, and their residing in dormitories and eating the food presented there. In another study, similar to our results, one-third of respondents claim to buy organic foods "very often" or "often", and another 43 percent claims that their purchase of such products is "rare" (Radman, 2005). In another study, rare buyers score highest (58.3%) while regular buyers score lowest (6.0%) (Tarkiainen & Sundqvist, 2005). Another research conducted with the young shows that they are not concerned about whether the food they consume is organic or not (71.8%), they attach importance to the taste (93%), price (78.7%), reliability (93.9%), healthiness (83.9%) as well as the appearance (75.3%) of the food (Bissonnette and Contento 2001). In a regular organic food consumer group, among the reasons behind their choices the most important for them is that organic foods do not contain chemicals, hormones or additives, while the reasons of taste, nutrition and price follow (Sarıkaya, 2007).

A higher educational level corresponds with more knowledge, a positive relationship between educational level with the knowledge and acceptance of organic food can be assumed (Stobbelaar et al., 2007). In the study conducted by Radman (2005), a positive correlation between education level and organic food consumption is detected. In this study, too, fourth-grade students' attitude scales are higher than the rest of the students. As people have more knowledge on a subject, they may develop better attitudes towards that subject. Considering the fact that the students participated are educated on health sciences, it can be stated that they develop positive attitudes towards these foods by evaluating them in terms of health.

Students did support organic food and stated that they would purchase more organic products if they were cheaper. Younger shoppers, in theory, support the organic industry, but this age group is unable to afford organic products (Essoussi & Zahaf, 2008). Similarly, our participants state that they would consume more organic products if they were cheaper or had similar prices with traditional foods. As stated above, the research group's consisting of students who are financially dependent on their families, namely their having less purchasing power, can be interpreted for their inability to buy organic products.

In the study carried out by Sarıkaya (2007), among the leading reason why consumers purchase organic products is environmental awareness. The study points out the lack of trust felt by consumers towards organic products. This situation also explains why organic products are not rapidly becoming widespread in Turkey. At the same time the study states that consumers make a point in obtaining information on control and supervision and think that non-organic products are harmful in terms of health. Consumers participating in the study consider it normal for organic product prices to be expensive and that they are ready to pay more compared to the alternatives to organic products. It has been determined that consumers have health concerns and for this reason are inclined to consume these products. The study points out that participants purchase these products mostly from supermarkets. While women compared to men give more importance to the supervision of organic products, women also find these products more reliable compared to men. As the level of education increases, the importance attached to environment friendly products also increases. It has also be noted that women pay attention to brand names when purchasing products.

Another study conducted in Turkey, they determined that factors, such as income, price, having information on organic foods, health and quality are also proven to influence purchasing behavior (Sanlıer et al., 2011).

Although organic food production and consumption increase, organic foods in Turkey are still in the introduction stage and not widely prevalent, thus resulting in low familiarity with them among the general public. In a study conducted to determine the potential demand for organic products in Ankara, Turkey, this demand is investigated based on four products with different price levels (tomato, cucumber, chicken meat, egg). Thirty three point four percent of the families that have filled in the questionnaire correctly know what an organic (natural and ecological) product is. Also, it comes to the fore that for the products in question (tomato, cucumber, chicken meat, egg) there is a consumer group willing to cover the price difference for organic products. According to these results, there is a considerable demand for the stated organic products (Koç et al., 2001). Consumers of organic foods are willing to pay more for their food to ensure the food has no growth hormones or that unnecessary chemicals and pesticides are not used during the production process (Bellows et al., 2008). As supported by the studies performed in these times in which the awareness about health gradually increases, there is a general demand for the production and consumption of organic foods. Although these products are comparatively expensive than other products, they have more demand and place in groceries as well as more organic-only stores. However, the consumption rate in our country is still lower than other countries. Therefore, these products should be made accessible to everyone spatially and financially by governmental regulations.

5. Conclusion

Environmental pollution appearing alongside with increasing technological progress also makes itself felt in the food chain. Especially, the widespread products with known or yet-unknown side effects for human health, such as genetically modified foods or foods with hormones, pose a serious threat to human health. For this reason, the demand for naturally produced organic products with no side effects increases day by day. As the education and income levels of the consumers rise, they have more awareness of and likewise show more demand for organic products (Sarıkaya, 2007). Nevertheless, despite the upturn in Turkey's domestic market, the percentage of organic agricultural production is still under 1%. Generally, consumers do not have enough knowledge about the features of organic products, such as their production techniques, their role in protecting the environment or health values (Sanlıer et al., 2011). By this study, the attitudes towards organic products of students who use them and the factors affecting their preferences are tried to be presented. In general, it is found out that the students' attitudes are positive but their buying rates are rather low.

The research shows that the average of the organic foods attitudes scale level of all 839 students is positive (150.2 ±21.6, out of 205 possible points). The points received from the "Positive statements regarding organic foods" section (42.3±7.42; out of 60 possible points) are lower than the points received from the "Negative statements regarding organic foods" (28.7±5.47, out of 40 possible points), and "Statements regarding organic food production" (14.5±3.29, out of 20 possible points), "Purchasing organic foods" (16.4±3.00, out of 20 possible points) "Comparing organic foods with traditional foods" (29.5±5.60, out of 40 possible points) and "Organic foods and chemical usage" (18.5±3.67, out of 25possible points) sections.

With the results obtained, we try to understand to what extend the students in the organic food market have awareness about organic products. These results are thought to be helpful in related companies' decisions concerning organic products.

6. Limitations of the study

This study has examined university students' attitudes towards organic foods; however, the results are limited only to Ankara University students who have participated in this study.

The results of this study reflect the attitudes and ideas of the student participants attending Ankara University Faculty of Health Sciences and cannot be generalized to other population frames.

7. Significance of the study

This study will allow researchers to better understand university students' attitudes towards organic foods. In addition, this research will allow researchers to better understand what health sciences students think when purchasing organic foods. University students are future buyers and will soon be making decisions on what food to purchase. Knowing how this group perceives organic foods will highlight areas for future research and education. Moreover, these students who are going to be working in health field will probably influence their consulters with their own perception of organic foods.

8. Appendix

Factor 1 Positive statements regarding organic foods	Factor loadings
20. Organic foods have better quality.	0.39
22. Organic foods are free of harmful effects.	0.37
29. I feel healthy when I eat organic foods.	0.60
31. I am protecting the environment by consuming organic foods.	0.42
36. I think organic foods are fresher.	0.49
38. I would buy organic foods as I see many benefits in them (for environment, health and animal rights).	0.54
42. If everyone consumed organic products, the world would be a better place.	0.62
46. Organically produced food means completely reliable food.	0.54
48. Infant foods with organic ingredients have more nutrients than traditional foods.	0.59
49. I live longer if I consume organic foods.	0.74
50. Consuming organic foods helps the prevention of obesity.	0.69
55. I think organic food consumption has no risks.	0.53
Eigenvalues:11.828 Variance explained by single factor:11.628 Alpha:0.86	
Factor 2 Negative statements regarding organic foods	
24. I am against my parents' or relatives' consumption of organic foods.	0.54
26. I do not consume organic foods because I do not want to try foods produced by new technologies.	0.54
40. Consuming organic foods are not amongst the actions I can take to protect the environment.	0.53
44. Organic infant foods are not as healthy as traditional foods.	0.54
45. Organic foods will bring harm to the society rather than benefits.	0.70
47. As it modifies the foods we consume, it is dangerous to eat organic foods.	0.68
51. I think the risks organic foods bring surpass their benefits.	0.64
52. Regular consumption of organic foods is harmful for my health.	0.69
Eigenvalues:2.902 Variance explained by single factor:21.194 Alpha:0.80	

Factor 3 Statements regarding organic food production	
14. Organic foods do not contain additives.	0.60
32. Organic foods do not contain preservatives or artificial colors.	0.69
33. Organic foods are produced without using hormones or antibiotics.	0.70
34. Organic foods are not genetically modified.	0.70
Eigenvalues:2.153 Variance explained by single factor:29.813 Alpha:0.85	
Factor 4 Purchasing organic foods	
13. I would be glad to consume more organic food as long as I can find them.	0.68
15. I think groceries should give more space to organic foods.	0.63
16. I think more agricultural land should be spared for organic foods.	0.67
21. Organic foods are more expensive.	0.54
Eigenvalues:1.732 Variance explained by single factor:37.558 Alpha:0.77	
Factor 5 Comparing organic foods with traditional foods	
6. Organic foods taste better than traditional foods.	0.53
7. Organic foods are more nutritious than traditional foods.	0.59
8. Organic foods reduce the risk of illnesses.	0.41
9. The vitamin and mineral content of organic foods are not more than those of traditional foods.	0.44
10. Organic foods are more reliable than traditional foods.	0.53
17. I do not think there is any difference between organic foods and traditional foods.	0.65
18. Organic foods do not have better quality than traditional foods.	0.67
23. I think the consumption of organic foods is a trend.	0.37
Eigenvalues:1.386 Variance explained by single factor:45.051 Alpha:0.79	
Factor 6 Organic foods and chemical usage	
1. Organic foods are not treated with any chemicals.	0.49
2. No chemical fertilizer is used in the production of organic foods.	0.60
3. Organic foods reduce soil contamination.	0.70
4. Organic foods reduce the use of pesticides in agriculture.	0.73
5. Organic foods secure the biological balance of nature.	0.62
Eigenvalues:1.319 Variance explained by single factor:51.999 Alpha:0.78	
Items that have been removed	
11. Organic foods look worse than traditional foods.	-
12. Organic foods have less shelf-life.	-
19. I do not buy organic foods as they are more expensive than traditional foods.	-
25. I do not consume organic foods because I cannot find them where I do my shopping.	-
27. I am not sure whether organic foods are really produced organically.	-
28. If I had thought they were really organic, I would have bought these foods.	-
30. Consuming organic foods means spending more money.	-
35. If organic foods were sold packaged and labeled, I would buy them.	-
37. I think that organic foods have better quality for they are sold in higher prices.	-
39. I think that traditional foods are as healthy as organic foods.	-

41. The companies producing organic foods are only after more profits.	-
43. Consuming organic foods is a political decision.	-
53. I think that organic food production is actively controlled by the government.	-
54. The government supervises the food industry in terms of organic foods.	-
56. I think that people are not informed enough about organic foods.	-
57. It takes time to get information about organic foods from the media.	-
58. An organic food can be presented to the market as a classical product.	-
59. Mandatory labeling regulation should not be preferred as it would increase the cost of organic foods.	-

The items and the factor loadings of this study

9. References

Ajzen, I. (1991). The theory of planned behavior. *Organizational Behavior and Human Decision Processes*, Vol. 50, pp. (179–211), ISSN: 0749-5978.

Anonymous, (2004). Current trends in the industry. Just - Foods. Retrieved from ABI/ global p. 10.

Anonymous, (2011a). Organik Tarım Nedir?. In: 09.08.2011, Available from: <http://www.tarim.gov.tr/uretim/Organik_Tarim,Organik_Tarim.html#EKOLOJ İK TARIM NEDİR >

Anonymous, (2011b). TKİB İzmir İl Kontrol Laboratuar Müdürlüğü Organik tarım Ürünleri ve Katkı Analizleri Laboratuvarı, In: 09.08.2011, Available from <http://www.izmir-kontrollab.gov.tr/organik.php >

Anonymous, (2011c). Genel Organik Tarımsal Üretim Verileri. In: 15.06.2011, Available from: <http://www.tarim.gov.tr/uretim/Organik_Tarim,Organik_Tarim_Statistikleri.html >

Anonymous, (2011d). Yıllara göre organik ürün ihracatımız. In: 15.06.2011, Available from: <http://www.tarim.gov.tr/uretim/Organik_Tarim,Organik_Tarim_Statistikleri.html >

Arvola, A., Vassallo, M., Dean, M., Lampila, P., Saba, A., Lahteenmaki, L., & Shepherd, R. (2008). Predicting intentions to purchase organic food: The role of affective and moral attitudes in the theory of planned behaviour. *Appetite,* Vol. 50, No. 2-3, pp. (443-454), ISSN: 0195-6663.

Atay, A., Sarı, E. 2007. *Organik Tarıma Başlarken* (E-book). Bursa Ticaret Odası, In: 09.08.2011, Available from http://www.keyifdunyasi.com/printArticle.php?aID=155).

Baker, S., Thompson, K., Engelken, J., (2004). Mapping the values driving organic food choice: Germany vs. the UK and UK vs. Germany. *European Journal of Marketing,* Vol. 38, No. 8, pp. 995 - 1012, ISSN: 0309-0566.

Bellows, A.C., Onyango, B., Diamond, A., & Hallman, W.K. (2008). Understanding consumer interest in organics: production values vs. purchasing behavior. *Journal of Agricultural & Food Industrial organization, Vol. 6,* pp. (1-27), ISSN: 1542-0485.

Bissonnette, M.M. Contento, I.R. (2001). Adolescents' perspectives and food choice behaviors in terms of the environmental impacts of food production practices: application of a psychosocial model. *Journal of Nutrition Education,*Vol. 33, No. 2, pp. (72-82), ISSN: 1499-4046.

Boccaletti, S. (2009). *Organic food Consumption: Results and Policy Implications,* Proceeding of Household Behavior and Environmental Policy", 3-4th June 2009, Paris.

Büyüköztürk, Ş. (2002). *Sosyal bilimler için veri analizi el kitabı* (Edition: 12th), Pegem Yayıncılık, ISBN: 9789756802748, Ankara.

Chen, M.F. (2007). Consumer attitudes and purchase intentions in relaiton to organic foods in Taiwan: Moderating effects of food-related personality traits. *Food Quality and Preference*, Vol. 18, pp. (1008-1021), ISSN: 0950-3293.

Chryssohoidis, G.M., Krystallis, A.2005. Organic consumers personal values research: Testing and validating the list of values (LOV) scale and implementing a value-based segmentation task. *Food Quality and Preferences*, Vol. 16, pp. (585-599), ISSN: 0950-3293.

Davies, A., Titterington, A.J. and Cocharane, C. (1995), Who buys organic foods? A profile of the purchasers of organic food in Northern Ireland, *British Food Journal*, Vol. 97, No. 10, pp. (17-23), ISSN: 0007/070X.

Essoussi, L. H. & Zahaf, M. (2008). Decision making process of community organic food consumers: an exploratory study. *Journal of Consumer Marketing*, Vol. 25, No. 2, pp. (95-104), ISSN:0736-3761.

Finch, J. E. (2005). The impact of personal consumption values and beliefs on organic food purchase behavior. *Journal of Food Products Marketing*, Vol. 11, No. 4, pp. (63-76), ISSN: 1045-4446.

Fotopoulos, C, & Krystallis, A. (2002). Organic product avoidance: Reasons for rejection and potential buyers' identification in a countrywide survey. *British Food Journal*, Vol. 104, No. 3/4/5, pp. (233-260), ISSN: 0007/070X.

Hughner, R. S., McDonagh, P., Prothero, A., Shultz II, C. J., & Stanton, J. (2007). Who are organic food consumers? A compilation and review of why people purchase organic food. *Journal of Consumer Behavior*, Vol. 6, pp. (94-110), ISSN: 1479-1830.

Kaya, H. G. (2003). Dünyada ve Türkiye'de Organik Tarımsal Ürün Ticareti ve Tüketici Reaksiyonları, In: 15.06.2011, Available from
< http://www.bahce.biz/organik/organik_ticareti.htm>

Kerlinger, F. N. (1973). Foundation of behavioral research. Second Edition, Hold, Rinehard and Winston, New York.

Kesici, T., & Kocabaş, Z. (1998). *Biyoistatistik*. Ankara Üniversitesi Ziraat Fakültesi Yayını No:75. Ankara Üniversitesi Basımevi, Ankara.

Kirazlar, N. (2001). Ekolojik (organic) Tarım Mevzuatı. *Proceeding of Türkiye 2. Ekolojik Tarım Sempozyumu*, Antalya, November, 2001.

Koç, A., Akyıl, N., Ertürk, Y. E., Kandemir, M. U. (2001). Türkiye'de Organik Ürün Talebi:Tüketicinin Kalite İçin Ödemeye Gönüllü Olduğu Fiyat Farkı. *Proceeding of Türkiye 2. Ekolojik Tarım Sempozyumu*, Antalya, November, 2001.

Lawrence, M. (2007). *College students' perceptions and Information sources regarding Organic and genetically Modified food industries*. Bachelor of Science in Agricultural Communications Oklahoma State University, pp:101, Stillwater, Oklahoma.

Lea, E., & Worsley, A. (2005). Australian consumers' food-related environmental beliefs and behaviours. *Appetite*, Vol. 50 No. 2-3, pp. (207-214), ISSN: 0195-6663/$.

Lockie, S., Lyons, K., Lawrence, G., & Grice, J. (2004). Choosing organics: A path analysis of factors underlying the selection of organic food among Australian consumers. *Appetite*, Vol. 43, No. 2, pp. (135-146), ISSN: 0195-6663.

Magnusson MK, Arvola A, Hursti U, Aberg L, Sjoden P. 2001. Attitudes towards organic foods among Swedish consumers. *British Food Journal*, Vol. 103, No. 3, pp. (209–227), ISSN: 0007/070X.

Magnusson, M. K., Arvola, A., Koivisto Hursti, U.-K., Aberg, L., & Sjoden, P. O. (2003). Choice of organic foods is related to perceived consequences for human health and to environmentally friendly behavior. *Appetite,* Vol. 40, pp. (109–117), ISSN: 0195-6663.

Makatouni, A. (2002). What motivates consumers to buy organic food in the UK? Results from a qualitative study. *British Food Journal,* Vol. 104, No. 3/4/5, pp. (345-352), ISSN: 0007/070X.

McEachern, M. G. & Willock, J. (2004). Producers and consumers of organic meat: A focus on attitudes and motivations. *British Food Journal,* Vol. 106, No. 7, pp. 534-552), ISSN: 0007/070X.

Onyango, B. M., Hallman, W. K., & Bellows, A. C. (2007). Purchasing organic food in US food systems. *British Food Journal,* Vol. 109 No. 5, pp. (399-411), ISSN: 0007/070X.

Padel, S., & Foster, C. (2005). Exploring the gap between attitudes and behaviour. *British Food Journal,* Vol. 107, No. B, pp. (606-625), ISSN: 0007/070X.

Radman, M. 2005. Consumer consumption and perception of organic products in Crotia. *British Food Journal,* Vol. 107, No. 4, pp. (263-273), ISSN: 0007/070X.

Roddy, G., Cowan, C. A. & Hutchinson, G. (1996). Consumer attitudes and behaviour to organic foods in Ireland, *Journal of International Consumer Marketing,* Vol. 9, No. 2, pp. (41-63), ISSN: 0896-1530.

Saba, A. Messina, F. 2003. Attitudes towards organic foods and risk/benefit perception associated with pesticides. *Food Quality and Preference,* Vol. 14, pp. (637-645), ISSN: 0950-3293.

Sanlıer, N., Kizanlikli, M. & Cöp, S. 2011. The determination of the perception, approach and behavior of teachers tqwards organic foods. *HealthMED,* Vol. 5, No. 2, pp. (307-316), ISSN:1840-2291.

Sarıkaya, N. 2007. Organik ürün tüketimini etkileyen faktörler ve tutumlar üzerine bir saha çalışması. Kocaeli Üniversitesi Sosyal Bilimler Enstitüsü Dergisi Vol. 14, pp. (110-125), ISSN: 1302-6658.

Schifferstein, H.N.J. & Oude Ophius, P.A.M. 1998. Health related determinants of organic food consumption in the Netherlands. *Food Quality and Preference,* Vol. 9, No. (3), pp. (119-133), ISSN: 0950-3293.

Shepherd, R., Magnusson, M., & Sjoden, P. (2005). Determinants of consumer behavior related to organic foods. *Ambio,* Vol. 34, No. (4/5), pp. (352-359), ISSN: 0044-7447.

SIPPO, 2011. *The organic market in Europe. Overview and market access information,* SIPPO and FIBL, Brunner AG, Druck und Medien, ISBN: 978-3-03736-186-3, Switzerland.

Solano, A. A. 2008. *Willingness to pay for organic and natural foods: do the Definitions of these terms affect consumer behavior?.* Masters of Science in Agricultural Economics. Faculty of the University of Delaware, pp.148, Delaware.

Storstad,O. & Bjørkhaug, H. 2003. Foundations of production and consumption of organic food in Norway: Common attitudes among farmers and consumers?. *Agriculture and Human Values,* Vol. 20; pp. (151-163), ISSN: 0889-048X.

Strobbelaar, D. J. Casimir, G., Borghuis, J., Marks, I., Meijer, L., & Zebeda, S. (2007). Adolescents' attitudes towards organic food: a survey of 15- to16-year old school children. *International Journal of Consumer Studies,* Vol. *31,* pp. (349-356), ISSN: 1470-6423.

Tabachnick, B.G., & Fidell, L.S.(1989). Using multivariate statistics. Harper Collins Publishing, USA.

Tarkiainen, A. & Sundqvist, S. 2005. Subjective norms, attitudes and intentions of Finnish consumers in buying organic food. *British Food Journal*, Vol. 107 No. (11), pp. (808-822), ISSN: 0007/070X.

The Nielsen Company. (2007). *Organic foods: A Canadian perspective.* New York, USA: The Nielsen Company. In: May 9, 2008, Available from < http://ca.nielsen.com/site/documents/OrganicFoodsMay2007.pdf>

Tregear, A., Dent, J. B., & McGregor, M. J. (1994). The demand for organically grown produce. *British Food Journal*, Vol. 96, pp. (21–25), ISSN: 0007/070X.

Tsakiridou, E., Boutsouki, C., Zotos, Y., & Mattas, K. (2008). Attitudes and behavior toward organic products: an exploratory study. *International Journal of Retail & Distribution Management,* Vol. *36,* No. 2, pp. (158-175), ISSN: 0959-0552.

Urena, F., Bernabeu, R., & Olmeda, M. (2008). Women, men and organic food: differences in their attitudes and willingness to pay. A Spanish case study. *International Journal of Consumer Studies,* Vol. 32 No. 1, pp. (18-26), ISSN: 1470-6423.

Vanderkloet, J.A.E. 2008. *Organic food consumers: Modeling the food choice process.* University of Alberta, Master of Science in Food Science and Technology, Department of Agricultural, Food and Nutritional Science, Edmonton, ISBN: 978-0-494-47430-3, Canada.

Vindigni, G., Janssen, M. A., & Jager, W. (2002). Organic food consumption: a multitheoretical framework of consumer decision making. *British Food Journal,* Vol. 104 No. 8, pp. (624-642), ISSN: 0007/070X.

Wandel, M., & Bugge, A. (1997). Environmental concern in consumer evaluation of food quality. *Food Quality and Preference,* Vol. 8, pp. (19–26), ISSN: 0950-3293.

Wen Chei Chan, F. 2008. *Values motivating the purchase of organic food: A laddering analysis.* University of Alberta. Master of Science in Food Science and Technology Department of Agricultural, Food and Nutritional Science, pp:102, Edmonton, ISBN: 978-0-494-47189-0, Canada.

Whole Foods Market. (2005). *Nearly Two-Thirds of Americans Have Tried Organic Foods and Beverages.* In: August 15, 2007, Available from http://www.wholefoodsmarket.com/pressroom/pr_11-18-05.html

Willer, H. 2011. *Agricultural land and other uses of certified organic areas. Research Institute of Organic Agriculture (FIBL).* Ackerstrasse, Switzerland European Organic Farming Statistics. The Organic-World.net homepage, FiBL, In: FiBL January 6, 2011. Available from < www.organicworld.net/statistics.html >

Zhao, X., Chambers, E., Matta, Z., Loughin, T. M., & Carey, E. E. (2007). Consumer sensory analysis of organically and conventionally grown vegetables. *Journal of Food Science,* Vol. 72 No. 2, pp. (S87-S91), ISSN: 0950-3293.

6

Do Consumers Pay Attention to the Organic Label When Shopping Organic Food in Italy?

Tiziana de Magistris and Azucena Gracia
Centro de Investigación y Tecnología Agroalimentaria de Aragón
Spain

1. Introduction

The main objective of organic agriculture is to produce healthy and quality food products without using synthetic chemical products in response to consumers' concerns about food safety, human health, animal welfare and the environment. The EU's promotion of organic agriculture constitutes an important alternative for society and for the more marginal agricultural regions in southern Europe.

First, organic agriculture provides consumers some products with higher nutritional value that are free of chemical agents. Second, it improves public health, and it brings significant benefits both to the economy as well as to the social cohesion of rural areas.

Organic label is a quality signal for the consumers and an important tool to help them to identify organic products. Yiridoe et al. (2005) stated that product labels can help buyers to assess product quality by transforming credence characteristics into search attributes. On 1 July 2010, the European Union introduced a new organic logo in order to increase the transparency and confidence in organic food products for consumers. The difference from the first logo, introduced in the 1990s, is its compulsory basis (EU n. 271/2010 of 24 March 2010) for those products in which at least 95% of the agricultural ingredients are organics. According to the European Union, the application of the new logo and labeling will lead to a much greater visibility of organic foods in the market and increase the trust among European citizens, who will be not only more informed but also more willing to purchase these products (Council Regulation (EC) No. 834/2007 of 28 June 2007).

Therefore, understanding the reason why consumers choose to buy organic products and investigating whether they pay attention to organic labeling when shopping are necessary in order to induce new individuals to buy organic food products and to increase the level of consumption among existing consumers.

The aim of our chapter is to analyze whether Italian consumers pay attention to organic labeling when shopping, and whether they are also affected by other psychological variables to make purchase decisions. We carried out a study on the intention to purchase organic food products, proposing a model of organic consumption which has been developed on the Theory of Planned Behavior by Ajzen (1991), which extends it with a new variable related to consumers' attention to organic label which has not been yet analyzed in Italy.

To achieve this goal, a survey was conducted with 380 food shoppers during November-December 2008 in Italy, and three-equation multivariate probit model is fitted, where consumers' attention to organic label, the intention to purchase, and organic consumption are the dependent variables. Findings show that paying attention to organic labeling influences the intention to purchase organic food products. Moreover, attitudes and lifestyles are the best predictors of organic purchase. However, environmental attitudes and subjective norms do not affect the intention to purchase them, while ethical dimension is considered one of main factors driving organic food purchases.

The rest of the chapter is structured as follows. Section 2 develops the theoretical framework where the relationship among endogenous and exogenous variables of our model are justified by literature review. Section 3 describes an overview of the European legislation on organic food labeling; Section 4 focuses on the data gathered and methodology. Section 5 shows the main results; and, finally, Section 6 concludes with a discussion of the marketing and policy implications.

2. Literature review

In general, research on the organic purchase intention is well documented in the academic literature (Saba and Messina, 2003; Tarkiainen and Sundqvist, 2005; Chen, 2007; Chen, 2008; Thogersen, 2007a; Dean et al., 2008; de Magistris and Gracia, 2009; Aertsen et al., 2009). Some studies analyzed the intention to purchase organic food products using the Theory of Planned Behavior (Ajzen, 1991) where the intention is considered the best predictor of behavior. In accordance with the TPB, the attitudes toward action sequence, as well as the relations among factors on the intention have been part of the general approach to attitudes in social psychology. A major theoretical position on attitudes and action has been formulated by Ajzen (1991) as the Theory of Planned Behavior (TPB), where the behavioural intention is basically determined by three factors: attitudes, subjective norms and perceived behavioural control. Attitudes are composed of behavioural beliefs and outcome evaluations of the consequences of beliefs. Subjective norms refer to perceived social pressure to perform or not perform the behaviour as perceived by the person. Finally, perceived behavioural control is the individual's beliefs about the amount of control that an individual has to successfully complete his/her behaviour (Ajzen, 1991).

Findings revealed that consumers' attitudes towards organic attributes (taste, health , food safety, environmentally friendly, animal welfare) and salient beliefs (attitudes towards its purchase, personal moral norms, ethical motivations and emotions) are the most important factors that explain consumers' decision-making process for organic food (Magnusson et al, 2001; Thøgersen, 2002; Saba and Messina, 2003; Millok et al., 2004; Verhoef, 2005; Tarkiainen and Sundqvist, 2005; Padel and Foster, 2005; Chryssohoidis and Krystallis, 2005; Honkanen et al., 2006; Thøgersen and Olander, 2006; Chen, 2007; Chen, 2008; Dean et al., 2008; Arvola et al., 2008; Aertsen et al., 2009; Gracia and de Magistris, 2008; de Magistris and Gracia, 2009; Guido et al., 2010). The results showed significant differences across the products and countries where the research was carried out.

2.1 Theoretical model

2.1.1 Attitudes

Attitudes towards the behavior refer to the degree to which an individual has a favorable or unfavorable evaluation of the behavior (Ajzen, 1991). According to Ajzen(1991) the more favorable the attitudes are towards a behavior, the stronger is the persons' intention to perform the behavior under consideration. Some studies about organic consumers reported

a significant positive relation between consumers' intention to purchase organic food and their attitudes toward organic food purchase (Saba and Messina, 2003; Millock et al., 2004; Chryssohoidis and Krystallis, 2005; Padel and Foster, 2005; Tarkiainen and Sundqvist, 2005; Honkanen et al., 2006; Thøgersen, 2007a; Thøgersen, 2007b ; de Magistris and Gracia ,2008)
Saba and Messina (2003) carried out their study in Italy, analyzing organic fruit and vegetable products. Findings showed that Italian consumers showed positive attitudes towards these products, perceived them as healthy, environmentally friendly, tastier and nutritious than conventional products. Millock et al. (2004) claimed that environmental and animal welfare attitudes influence organic food choice to a lesser extent than attitudes towards taste, freshness and health aspects of organic food. Chryssohoidis and Krystallis (2005) indicated that the most important motives behind the purchase of organic products in Greece are healthiness and better taste of the organic food. However, environmental motives influence organic food choice to a lesser extent. Padel and Foster (2005) concluded that consumers buy organic food products because they perceived them to be better for their health. Moreover, they found that the attitude towards environmental protection was also a factor that explains organic food buying decisions.

Tarkiainen and Sundqvist (2005) analyzed the factors that affected the purchase of organic food in Finland. The results suggested that consumers 'intentions to buy organic food are predicted by their attitudes and environmental concerns. Honkanen et al. (2006) studied the motives driving organic food choice in Norwegian consumers. The findings indicated that ecological motives and attitudes towards organic food have a significant influence on the intention to purchase organic food. Similar results were found by Thøgersen (2007b) who stated that health, taste and environmental consequences were the most important factors related to organic consumers' attitudes toward buying organic tomato juice. Along the same line, de Magistris and Gracia (2008) investigated the organic purchase behavior of Italian consumers. They found that consumers' attitudes towards organic food, in particular towards the health attributes and towards the environment are the most important factors explaining organic food purchase

2.1.2 Subjective norms

Ajzen (1991) identified also the role of subjective norms, which refer to the perceived social pressure to perform or not to perform the behavior. In other words, subjective norm is an individual's conviction that acting in a certain way is right or wrong regardless of personal or social consequences. Few studies, however, have examined subjective norms in relation to organic food purchases.

Sparks and Shepherd (1992) included subjective norms in their study of green consumers, but their explanatory power was relatively weak, even though significant. Tarkiainen and Sundqvist (2005)) investigated the correlation between subjective norms and attitudes towards organic food. The authors suggested that the link could be explained by the information that individuals have towards social environment. However, the authors did not find direct significant relation between subjective norms and the intention to buy organic foods. On the other hand, Thøgersen (2007b), and Dean et al. (2008) found a significant positive relation between consumers' intention to purchase organic food and their subjective norms. Finally, Smith and Paladino (2010) in their study of organic consumers in Australia stated that subjective norms are significant factors influencing the decision-making of consumers.

2.1.3 Perceived behavioural control

Perceived control is the "perception of the ease or difficulty of performing the behavior of interest anticipating some obstacles" (Ajzen, 1991). In the context of organic consumer behavior, price and availability potentially limit organic purchases. The past research of organic consumption has shown that the most important reasons for not buying organic food are lack of availability and organic food's relatively higher price compared to conventional food products. (Magnusson et al., 2001; Fotopoulos and Krystallis, 2002; Zanoli and Naspetti, 2002; Padel and Foster, 2005; Chryssohoidis and Krystallis, 2005). However, Canavari et al. (2002) found that 30 per cent of consumers are willing to pay a price premium directly to farmers.

2.1.4 Ethical dimension

In literature on organic consumer behavior, major extension of the TPB is related to the ethical dimension, defined by Shaw and Clarke (1999) as "the degree to which consumers prioritize their own ethical concerns when making products choices." Some studies used different variables in order to take into account ethical dimension in organic consumer behavior. To illustrate, Michaelidou and Hassan (2008) used self-identify concept defined as the enduring characteristics that people ascribe to themselves. The authors found that self-identity affected the intention to buy organic food products. Honkanen et al. (2006) mentioned ethical values based on environmental and animal protection in order to explain organic behavior, while Arvola et al. (2008) applied moral norms, defined as an individual's conviction that acting in a certain way is right or wrong regardless of their personal or social consequences, to demonstrate their strong influence on the consumption of organic food in Italy and in the United Kingdom.

2.1.5 Socio-demographic variables and lifestyles

Socio-demographic characteristics were found to be significant in explaining the decision to buy organic foods mainly in empirical studies conducted in Europe. Findings revealed that only sex, income, age, education and household size were significant (Canavari et al., 2002; Millock et al. 2003; Lockie et al. 2004; and Tsakiridou et al. 2006; Gracia and de Magistris, 2008). To illustrate, older, more educated consumers with high income and those living in larger households are more likely to buy organic food products.

The concept of "food-related lifestyle" was developed in the mid-1990s as a segmentation tool, which is tailored to the consumer's role as a food shopper (Grunert and Brunsø, 1997). Lifestyles are important exogenous factors in the decision-making process influencing consumer behaviour (Gil et al, 2000; Brunsø, Scholderer and Grunert, 2004). Schifferstein and Oude Ophuis (1998) stated that health behaviour, such as additional exercise and habits related to food intake, affects positively the organic food choice and Chryssohoidis and Krystallis (2005) stated that most of Greek organic buyers claimed to follow a balanced and healthy diet. de Magistris and Gracia (2009) demonstrated that lifestyles represent one of main factor explaining organic food decision-making. In particular, organic food products are preferred by those consumers interested in maintaining a particular lifestyle associated with healthy and eating habits

2.1.6 The extension of planned behaviour model: paying attention on organic labelling and organic knowledge

Academic studies of organic food labels have increased substantially in Europe during the last two decades. Most of the empirical works have used the hypothetical approach to assess consumer preferences for organic food labels to calculate consumers' Willingness to Pay (WTP) for the presence of them, implying their intention to purchase organic foods, taking into account the presence of organic label on packaging. Paying attention to organic labels depends also to the degree of self-knowledge that consumers have towards organic food products when shopping. Moreover, consumers might show a certain level of knowledge about the information that organic label provides to them such as that organic food products, free OGMs, they support the animal welfare and they content 95% of organic ingredient at least. Hence, organic food label perceived by consumers is an important issue in the organic food market because it represents the only instrument that consumers have to differentiate the attributes of organic food products from those of conventional ones, and build positive attitudes towards organic food (von Alvesleben, 1997). Moreover, Yiridoe et al. (2005), in their literature review, stated that knowledge of organic food products can affect consumers' organic buying decision for two reasons. The first one, the lack of knowledge of organic labeling is considered the number one reason why consumers do not buy organic food. The second reason is that consumers who do not consider that organic food products have enough detailed information cannot clearly differentiate the unique attributes of organic from conventionally grown alternatives. Usually, many organic food consumers identify organic products based on the organic logos and labels attached to the product. The authors concluded that information about organic food helps to transform the credence characteristic of such products into search attributes, thereby allowing the consumer to better evaluate the quality before deciding to buy the product. Organic knowledge can be considered as a part of formation that consumers acquire from organic food products. Hill and Lynchehaun (2002) found that knowledge represents an important factor influencing the purchase of organic products. Poelman et al. (2008) analyzed whether information on organic production and fair trade affects the preference for and perception of pineapples in British and Dutch individuals. The results indicated a slight positive impact of organic product information on consumers' preference and perception for pineapple. Gracia and de Magistris (2008) and de Magistris and Gracia (2009) also provided evidence on the positive influence of consumers' organic knowledge on purchasing behavior. Based on these previous findings, the model of intention to purchase organic food products is presented in figure 3.

3. EU organic labeling legislation

The EU legislation on organic labeling is laid down in the following four Regulations: i) Regulation (EEC) No 2092/21, ii) Commission Regulation (EC) No 331/2000 , iii) Council Regulation (EC) No. 834/2007 of 28 June 2007 and, iv) Commission Regulation (EU) No 271/2010.

Regulation (EEC) n. 2092/91 was adopted in 1992 when organic agriculture received the official recognition from EU members within the EU Common Agricultural Policy. The main importance of this Regulation is it created the common minimum standards for the entire European Union inducing European citizens to trust organic products coming from other European countries. Even though plant products were regulated, additional rules were laid down later. These rules regarded animal feed, prevention of illness, veterinary treatment, animal protection, and livestock breeding. Moreover, the use of genetically modified organisms was expressly prohibited while the imports of organic products from Third Countries were allowed after being recognized as equivalent by EU.

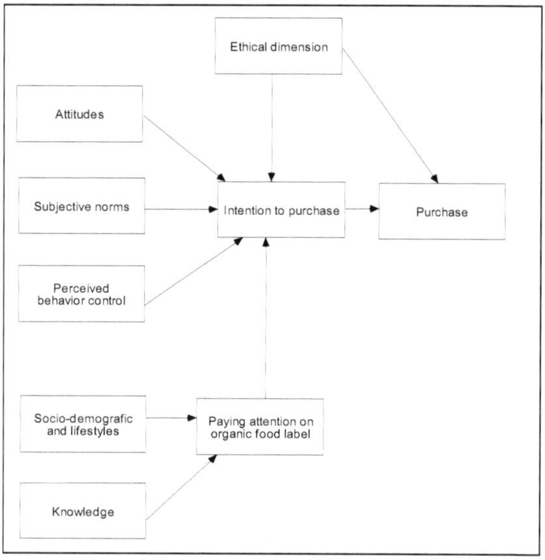

Fig. 3. Model of intention to purchase organic food products

The first European organic logo was introduced in March 2000 by Commission Regulation (EC) No 331/2000 bearing the words "Organic Farming-EC control systems" This logo had voluntary basis for those producers who satisfied the Regulation (EEC) No 2092/91.
The indicator that a product was covered by the inspection scheme had to be shown in the same language or languages as used for the labeling, as shown in figure 1.
Moreover, the Regulation authorized the combination of two indicators referring to the national languages, and the minimum size for a logo with a single indicator was 20mm diameter while for a logo with a combination of two indicators the minimum size was 40mm diameter.

Fig. 1. Organic logo introduced in Regulation (EEC) No 2092/91.

The Regulation (EC) n. 834/2007 was laid down on June 2007, and it revises of the Regulation No 2092/91. The need to review it was prompted during a public conference titled "Is Organic farming: ready for the next decade?" in December 2006, which represented an opportunity for the 90 stakeholders in attendance to discuss a possible proposal of revision of the Organic Regulation at the European level.
The main elements of the Regulation (EC) n. 834/2007 were a new definition of organic productions; the application of organic production to aquaculture, seaweeds, wine making

and yeast; the main labeling provisions; new rules on inspection; import and labeling, and its applicability after January 1, 2009.

According to the art. 1, organic production was defined as an overall system of farm management and food production whose function was to combine the best environmental practices, a high level of biodiversity, the preservation of natural resources, the application of animal welfare standards in order to meet consumers' demand with products of high quality. The Regulation dedicated a complete section to organic labeling rules. To illustrate, all usual common terms for "organic" in the different member states and its derivatives and diminutives were equally protected in their use. The term "organic" was not applied to agricultural products which clearly had no connection with organic production and those which contained GMOs (Genetic Modified Organisms). Nevertheless, the organic labeling had to explicate the general limit of 0.9 % for the accidental presence of GMOs. Secondly, the new European logo became compulsory on packaged European products, even though national and private logos could still be used. For the first time, the organic logo was compulsory for only those food products with at least 95% of the organic ingredients whose the raw materials composed the products. The indication had to mention i) EU Agriculture where the agricultural raw material had been farmed in the EU., ii) non-EU Agriculture, where the agricultural raw material came from third world countries, and iii) EU/ non-EU Agriculture, where the origin of raw material was mixed, meaning that it had been farmed both in the UE countries and third World countries. Organic ingredients in non-organic food had to be listed as organic in the list of ingredients, while the code number of the control body had to be indicated in order to guarantee better transparency.

Since the introduction of a new logo was supported by all European States, the European Commission organized an EU-wide organic logo competition opened to art and design students from all EU countries. The competition was won by the German design student Dušan Milenković, who received 63% of the overall votes from the European public. Hence, European Union launched a second certifying symbol in the Regulation 271/2010 where the organic logo was definitively mandatory, also called: "Euro-leaf" which symbolized the marriage of Europe (the stars derived from the European flag) and Nature (the stylized leaf and the green color), as shown in figure 2. In accordance with this last regulation, there will be a two year transition period for allowing its introduction on the packaging.

Fig. 2. Organic logo introduced in Commission Regulation 271/2010

In Italy, the certification bodies, which carry out the organic certification, are authorized by the Ministry of Agriculture. They are private entities and their main task is to ensure compliance with the implementation of regulations by the organic farms and give their own brand labels to be affixed to products sold. To illustrate, certification body has own brand label which has to be used only by those organic farmers associated to it in order to avoid that one organic farmer achieves organic certification from more than certification enterprise. In general, they do not have to entertain any relationship with organic farms and they carry out inspections every year. The inspection consists in compliance with

regulations and collecting some sample for testing at ARPAT or at a laboratory accredited Sinal (Sistema Nazionale per l'Accreditamento di Laboratori).
In Italy, there are 17 certification bodies recognized and their activity is constantly under control of the Region where they are placed. Moreover, 14 certification bodies over 17 are located in the Centre and North of Italy, whereas only two of them carry out their activity in Southern Italy. In particular, BIKO, IMO Gmbh and Q.C. & I. have been recognized to operate in Bolzano province.

ABCERT Srl - Cod. Min. IT BIO 013	Terlano (BZ) Web: http://www.abcert.it
BIKO - Kontrollservice Tirol (Cod. Min. IT - BIO - 001 - BZ)	INNSBRUCK - Austria Web: http://www.biko.a
Bioagricert S.r.l. - Cod. Min. IT BIO 007 - (ex BAC) E-mail: info@bioagricert.org	Casalecchio di Reno BO Web: http://www.bioagricert.org
BIOS S.r.l. - Cod. Min. IT BIO 005 - (ex BSI)	Marostica VI Web: http://www.certbios.it
BIOZOO - S.r.l. - Cod. Min. IT BIO 010 - (ex BZO)	Sassari SS Web: http://www.biozoo.org
CCPB S.r.l. - Cod. Min. IT BIO 009 - (ex CPB)	Bologna BO Web: http://www.ccpb.it/
CODEX S.r.l. - Cod. Min. IT BIO 002 - (ex CDX)	Scordia CT Web: http://www.codexsrl.it/
EcoGruppo Italia S.r.l. - Cod. Min. IT BIO 008 - (ex ECO)	Catania CT Web: http://www.ecogruppoitalia.it
ICEA - Istituto per la Certificazione Etica e Ambientale - Cod. Min. IT BIO 006 - (ex ICA)	Bologna Web: http://www.icea.info/
IMC - Istituto Mediterraneo di Certificazione S.r.l. - Cod. Min. IT BIO 003 - (ex IMC)	Senigallia (AN) Web: http://www.imcert.it/
IMO Gmbh (Cod. Min. IT - BIO - 002 - BZ)	8570 Weinfelden Web: http://www.imo.ch
Q.C. & I. - Gesellschaft für kontrolle und zertifizierung von Qualitätssicherungssystemen GMBH (Cod. Min. IT - BIO - 003 - BZ)	Tiergartenstr. 32 D- 54595 Prum Web: http://www.qci.de
QC S.r.l.- Cod. Min. IT BIO 014	Monteriggioni SI Web: http://www.qcsrl.it/
Sidel S.p.a - Cod. Min. IT BIO 012 - (ex SDL)	BOLOGNA BO Sidel S.p.a - Cod. Min. IT BIO 012 - (ex SDL) Web: http://www.bio.sidelitalia.it/
Suolo e Salute - Direzione Tecnica e Ufficio Estero - Cod. Min. IT BIO 004	Bologna (BO) Web: http://new.suoloesalute.it/
Suolo e Salute srl - (Cod. Min. IT - ASS	Roseto degli Abruzzi (TE) Web: http://www.suoloesalute.it
Suolo e Salute srl - (Cod. Min. IT 004 - ex ASS)	Taormina (ME) Web: http://www.suoloesalute.it

Table 1. Certification bodies recognized in Italy.

4. Material and methods

4.1 Multivariate probit model

The three endogenous variables of the above intention to purchase model defined in figure 3 (organic food purchase, the intention to purchase and paying attention to organic food label) are discrete variables. The first equation in the model is consumers' organic food purchase (OP), specified as follows:

$$OP_i^* = \lambda IP_i^* + \beta X_i + u_i \tag{1}$$

Where IP_i^* is the variable related to the intention to purchase organic food products defined below, X_i is a vector of exogenous variables related with the ethical dimension of organic purchase, and u_i is the error term normally distributed N(0, σ_u^2). OP_i^* is unobserved; what is observed is the purchase of organic food products stated by the individual when shopping, where OP_i^* =3 means that a consumer reports that he usually purchases organic food, IPi = 2 means that the consumer sometime buys organic food and IPi = 1 means the consumer never has bought them. Observed and latent variables are related as follows:

$$\begin{aligned} OP_i &= 1 \quad if \quad OP_i^* \le \tau_1 \\ OP_i &= 2 \quad if \quad \tau_1 \le OP_i^* \le \tau_2 \\ OP_i &= 3 \quad if \quad \tau_2 \le OP_i^* \le \tau_3 \end{aligned} \tag{2}$$

where τ_i are the unknown threshold parameters to be estimated. The first threshold parameter is normalized to zero (τ_1 = 0).

The intention to purchase organic food products (IP) equation is defined as follows:

$$IP_i^* = \lambda ORG\_LABEL_i^* + \beta Y_i + \varepsilon_i \tag{3}$$

where $ORG\_LABEL_i^*$ is the variable related to consumers' paying attention to organic labels when shopping defined below, X_i is a vector of all exogenous variables (attitudes, subjective norms, perceived behavioural control, and ethical dimension), and ε_i is the error term normally distributed N(0, σ_u^2). IP_i^* is an unobserved variable, which reflects the intention to purchase organic food products, stated by the individual when shopping and defined in 2 levels, as follows:

$$\begin{aligned} IP_i &= 1 \quad if \quad IP_i^* \ge 0 \\ IP_i &= 0 \quad if \quad IP_i^*, \le 0 \end{aligned} \tag{4}$$

The level of paying attention to organic labels is defined as:

$$ORG\_LABEL_i^* = \varpi Z_i + \xi_i \tag{5}$$

where, Y_i represents all the exogenous variables (socio-demographic, lifestyles and, knowledge) and ξ_i is the normally distributed error term N(0, σ_ξ^2). ORG_LABEL_i* is the unobserved paying attention to organic label, and as before, the researchers observe an ordered level of paying attention, as follows:

$$
\begin{aligned}
&ORG\_LABEL_i = 1 \quad if \quad ORG\_LABEL_i^* \leq \upsilon_1 \\
&ORG\_LABEL_i = 2 \quad if \quad \upsilon_1 \leq ORG\_LABEL_i^* \leq \upsilon_2 \\
&ORG\_LABEL_i = 3 \quad if \quad \upsilon_2 \leq ORG\_LABEL_i^* \leq \upsilon_3 \\
&ORG\_LABEL_i = 4 \quad if \quad \upsilon_3 \leq ORG\_LABEL_i^* \leq \upsilon_4 \\
&ORG\_LABEL_j = 5 \quad if \quad \upsilon_5 \leq ORG\_LABEL_i^*
\end{aligned}
\qquad (6)
$$

where υ_I are the unknown threshold parameters to be estimated. The first threshold parameter is normalized to zero ($\upsilon_1 = 0$).
To estimate the three equations [1], [3] and [5], we have assumed that the error terms (u_i, ε_i and ζ_i) may be correlated. The, instead of estimating them independently, they are considered to be a multivariate limited-dependent-variable model, in which the three error terms (u_i ε_i and ζ_i) follow a multivariate normal distribution with mean zero and variance and covariance matrix Ω. The limited dependent-variable (LDV) model with correlated error terms are estimated using Hajivassiliou and McFadden's (1998) procedure implemented in Proc QLIM in the SAS© 9.1 statistical software package.

4.2 Data gathering and variables definitions

Data were collected from a survey conducted in Italy during November-December 2008. The capital Rome was chosen for two reasons. The city can be considered a representative sample of Italy since its economic indicators, such as average expenditure on food products (498 €), average household income per year (34,000 €) and most demographic characteristics are very close to the economic indicators for Italy (ISTAT, 2008). Sample size was set at 380. As the population can be considered infinite, this sample size results in a sampling error of ±5%, assuming a confidence level of 95.5% (k=2) and p=0.5. A stratified random sample of consumers was made on the basis of town district and age. A number of representative grocery stores and supermarkets were selected in each town district, and food shoppers were randomly selected outside these food outlets. Target respondents were the primary food buyers in the household, and interviews were carried out face to face. Interviewers approached the randomly selected individuals asking them one screening question, whether they were the main household food shopper. The questionnaire was designed to analyze consumers' organic food knowledge, attitudes and purchase behavior.
The questionnaire format was validated using a pilot survey. The questionnaire also contained questions on socio-demographic characteristics (i.e. sex, family size and composition, age, education level, income) and lifestyles. Summary statistics for the characteristics of the full sample are presented in Table 2. The majority of participants are female (58%) and, on average, participants are living in households of 3 members. In addition, the average age is about 45 years. Moreover, around 15 % belong to high income groups and about 48% of the subjects have a university degree.
The first question was related to their knowledge of organic food products. The level of self-reported knowledge was measured asking respondent their self-reported level of knowledge from 1 to 3, where 3 indicate the highest level of knowledge (KNOW). Objective

knowledge was measured by four dummy variables (GMOs, NATURAL, ANIMAL WELFARE and, 95_INGREDIENTS) which take 1 value if respondent states that organic food products are GMOs free, natural, they support the animal welfare and, they content 95% of organic ingredients at least, and 0 otherwise (see table 3)

Variable definition	Name (type)	Value %
Gender		
Male	FEMALE (dummy)	42
Female		58
Household Size	HSIZE (continuous)	2.9
Age of respondent (average)	AGE	45
Education of respondent		
Elementary		8.1
Secondary		43.5
University	UNIVERSITY (dummy)	48.4
Household Income:		
Net income lower than 1,500 €/month	LW_INC (dummy)	23.1
Net income between 1,500 and 2,500 €/month		63.1
Net income higher than 2,500 €/month		14.8

Table 2. Sample characteristics

Exogenous variables	Name (type)	Value
Self-reported knowledge on organic foods		
High (3)	KNOW	44.4
Medium (2)		48.42
Low (1)		7.11
Objective knowledge on organic foods		
They are free GMOs (dummy)	GMOs	78%
They are natural (dummy)	NATURAL	63%
They support the animal welfare (dummy)	ANIMAL WELFARE	72%
They content 95% of organic ingredients at least (dummy)	95_INGREDIENTS	77%

Table 3. Self-reported knowledge and organic knowledge

The second set of questions were related to organic food consumption (consumption level, intention to purchase, frequency of purchase, perceived quality, place of purchase, etc.). Finally, several questions about consumers' attitudes towards organic food products and environmental aspects were included.

Table 4 presents the definition of the endogenous variables. More than 50% of consumers declare to be an habitual buyer of organic food products; and 82% of them are willingness to buy them (82%). Moreover, around 59% of Italian consumers state that probably yes or definitely yes they pay attention to organic label when shopping organic food products.

Variable definition Endogenous variables	Sources	Name (type)	Value
Purchase organic food products Habitual=3 Occasion=2 Never=1	Chen (2008)	OP	50.7 28.7 20.5
Intention to purchase organic food products Yes =1 No=0	Chen (2008)	IP	81.9 18.0
Do you pay attention on organic label when shopping? Definitely no (1) Probably No (2) Indifferent (3) Probably Yes (4) Definitely yes (5)	Chen (2008)	ORG_LABEL	6.3 9.5 25.5 36.8 21.6

Table 4. Endogenous variables definition

The definition of the psychological variables used in the model has been done based on previous empirical papers (Table 5). Respondents were asked to indicate their agreement or disagreement with the statements provided using a five point Likert scale where one indicates strong disagreement and five, strong agreement. The scale items for the different aspects and the empirical papers used are shown in table 5, while lifestyles are showed in table 6.

Exogenous variables	Source	Name(type)	Value
Attitudes towards organic food products			
Organic food products are healthier	Michaelidou and Hassan (2010) Gil et al. (2000) Chen (2007)	HEALTH	3.7
Organic food products have superior quality		QUALITY	4.1
Organic food products are safer		SAFETY	3.9
Organic food products seem to be fresher than conventional ones		FRESH	3.5
Organic food products are the same as conventional ones		SAME	2.9
Organic food products are more expensive		EXPENSIVE	3.9
Organic food products are environmental friendly		ENVIRON	3.8
Organic food products are in fashion		FASHION	3.0

Exogenous variables	Source	Name(type)	Value
Attitudes towards organic food purchase			
I believe that buying organic food products is good	Chen (2007) Chen (2008)	GOOD	3.9
I really support organic purchase		SUPPORT	4.1
Subjective norm			
Most people who are important for me think that I should buy organic food products	Bredahl (2001) Chen (2007) Chen (2008)	SNORM	3.6
Perceived behavioral control			
Whether I will eventually buy organic food products is entirely up to me	Bredahl (2001) Chen (2007) Chen (2008)	CONTROL	3.5
If organic food products were available in the shops, I do not think I would ever be able to do so		DIFFICULTY	2.8
Ethical dimension	Arvola et al. (2009) Thøgersen (2002)		
I feel I ought to choose organic food products		MORAL	3.1
I fell obligation to choose organic food products		PROMOTION	2.3
I feel I should choose organic food products		FEELSHOULD	2.7

Table 5. Psychological variable definition

Exogenous variables Lifestyles		Name(type)	Value
Consumer do exercise		EXCISE	51.6
Consumer avoid snaking during between lunch and dinner		NOSNACK	3.17
Consumer avoid to be stressed		STRESS	3.43

Table 6. Lifestyles

5. Results

The estimated parameters for the model defined by [1], [3] and [5] equations, using variables in table 2,3,4,5,and 6 are presented in table 7. Only exogenous variables statistically different from zero, at a significant level of 0.05, have been finally included. Results show that correlations between equations are significant at 5% level. It means that errors for all equations are correlated. Therefore, we can conclude that the equations are not independent and that the simultaneous estimation of both equations in the model is the appropriate approach to obtain consistent parameter estimates.

The organic label equation (ORG_LABEL) includes socio-demographic characteristics, self-reported knowledge, objective knowledge, and lifestyles (FEMALE, LW_INC, KNOW, GMOs, NOSNACK and STRESS). FEMALE has a positive significant effect on paying attention to organic label, while LW_INC is negative. Findings are in accordance with Canavari et al. (2002); Millock et al. (2003); Lockie et al. (2004), Tsakiridou et al.(2006); Gracia and de Magistris (2008). Moreover, the positive and statistically significant estimated coefficient for the variable KNOW indicates that, if consumers show a high degree of self-reported organic knowledge, they are more likely to pay attention on organic label when shopping. However, only the variable GMOs related with organic knowledge has been found with a positive sign and it is statistically significant.

On the other hand, as we expected from other studies (Schifferstein and Oude Ophuis, 1998; Chryssohoidis and Krystallis, 2005; de Magistris and Gracia, 2009) variables related to lifestyles (NOSNACK and STRESS) are statistically significant, meaning that those consumers who usually do not have snack and they try not to stress are more likely to pay attention on organic label when shopping.

To pay attention to organic food label (ORG_LABEL) has been statistically significant on the intention to purchase (IP) equation. The positive coefficient associated with the ORG_LABEL variable in the intention to purchase equation indicates that the consumers who pay attention to the organic labels are more likely to buy organic food products. As we expected there is a significant relation between the intention to purchase organic food products (IP) and other variables related with attitudes towards purchase (GOOD). In addition, the positive and statistically significant estimate coefficients for the HEALTH and FRESH variables indicate that the more consumers believe that organic foods are healthier and fresher than conventional food ones, the more likely they are to buy organic food products. Finally, there is a positive and significant relation between the intention to purchase organic food products and the perceived behavior control (DIFFICULTY). These findings are consistent with those reported in Magnusson et al. (2001), Fotopoulos and Krystallis (2002), Zanoli and Naspetti (2002), Padel and Foster (2005) Chryssohoidis and Krystallis (2005), Tarkianen and Sundqvist (2005), Honkanen et al. (2006) and, Thogersen (2007). However, subjective norms (SNORM) and moral norms (MORAL) have not found statistically significant from zero. Finally, the intention to purchase organic food (IP) estimated parameter is positive and statistically significant in final organic purchase decision (OP). This means that as Ajzen postulates, the intention to purchase is the best predictor of final behavior, in our case organic food purchase. Only one additional variable, moral norm (MORAL) positively influences the final purchase decision.

We calculated the marginal effects to assess the effects of the exogenous variables on ORG_LABEL, IP, and OP variables (ordinal variables). In this specific case, and for the continuous exogenous variables, effects are calculated by means of the partial derivatives of the probabilities with respect to a given exogenous variable. In the case of dummy variables, the marginal effects are calculated taking the difference between the predicted probabilities in the respective variables of interest, changing from 0 to 1 and holding the rest constant. The change in predicted probabilities gives a more accurate description of the marginal effect of a dummy variable on event probability, than predicting the probability at the mean level of the dummy variable. The marginal effects for the continuous variables and for the dummy variables are shown in Tables 8, 9 and 10.

Coefficients	ORG_LABEL Estimates	t-ratio		IP Estimates	t-ratio		OP Estimates	t-ratio	
Intercept	-0.798	-2.21	**	0.44	2.88	**	1.57	6.76	**
FEMALE	0.29	2.24	**						
LW_INC	-0.543	-3.26	***						
KNOW	0.202	1.91	*						
GMOs	0.55	3.32	**						
NOSNACK	0.267	4.33	**						
STRESS	0.224	3.60	**						
ORG_LABEL				0.44	4.04	**			
GOOD				0.37	4.05	**			
HEALTH				0.29	2.90	**			
FRESH				0.20	1.71	**			
DIFFICULTY				-0.25	-2.22	**			
IP							0.31	3.82	**
MORAL							0.312	3.68	**
N	380								**
Log Likelihood	-704.2								**
μ_2	0.60	5.55	**				0.90	9.92	**
μ_3	1.49	11.19	**						
$\mu 4$	2.66	17.14							
Correlations									
Organic label paying attention				0.32	1.91	*	0.16	2.05	**
Intention to purchase							0.75	7.59	**

(**) (*) denotes statistical significance at the 1 (5) (10) per cent significance levels

Table 7. Results of Multivariate probit model

With respect to the marginal effects on consumers' pay attention to organic label when shopping (table 8), results indicate that female consumers have a higher probability to pay attention to organic label. Moreover, as we expected, there is a negative relationship between low income and paying attention to organic label. One of possible explication of these results is that people with low income are more likely to have a lower attention to organic labeling when shopping because of the higher price of the products.

The marginal effects of self-reported and objective knowledge are as expected. Those consumers who believe that organic food label is GMOs free, and they state a higher self-reported knowledge of these products are more likely to state a higher probability to pay

attention to organic label when shopping. Finally, those consumers who strongly agree with not snaking and avoiding stress life are more likely to pay attention to organic food labels.
Regarding the intention to purchase organic food products, the results indicate that consumers with higher probability to pay attention to organic labels (ORG_LABEL) are more likely to be willing to buy organic food products (table 9). As consumers present more positive attitudes towards the purchase (GOOD) they are more likely to buy them. Moreover, the more subjects consider difficult to find organic food products where they usually shop, the less likely to be willing to buy them. As we expected, those consumers believe that organic food products are fresher and healthier (FRESH, HEALTH) than conventional ones, they are more likely to purchase them. Finally, with respect to the final purchase of organic food (table 10), findings suggest, as expected, that consumer who strongly agree that it is moral to buy organic food products (MORAL) they are more likely to purchase them. Finally, as we expected the marginal effects of the intention to purchase organic food products are positive. Those consumers who strongly state the intention to purchase organic food products are more likely to finally purchase them.

	Prob ORG_LABEL =1	Prob ORG_LABEL =2	Prob ORG_LABEL =3	Prob ORG_LABEL =4	Prob ORG_LABEL =5
Variable					
FEMALE	-0.028**	-0.038**	-0.052**	0.0407**	0.077***
LW_INC	0.057**	0.066**	0.074**	-0.083**	-0.112***
KNOW	-0.017**	-0.025**	-0.037**	0.030***	0.055***
GMOs	-0.051**	-0.052**	-0.064**	0.065**	0.091***
NOSNACK	-0.023**	-0.033**	-0.049**	0.033**	0.073***
STRESS	-0.019**	-0.027**	-0.040**	0.027**	0.059**

Table 8. Marginal effects of paying attention to organic labeling

	Prob Intention to purchase=1
Variable	
ORG_LABEL	0.062***
GOOD	0.078***
HEALTH	0.052**
FRESH	0.039***
DIFFICULTY	-0.047**

Table 9. Marginal effects of the intention to purchase organic food products

	Prob OP=1	Prob OP=2	Prob OP=3
Variable			
IP	-0.349***	0.167**	0.182***
MORAL	-0.105***	0.037**	0.678***

Table 10. Marginal effects of organic food purchase

6. Final remarks

In 2010 a new organic labelling was introduced in the European Union in order to increase the trust among European consumers towards organic food products and communicate them more environmental signals. The new label is compulsory and used only in those products which contain at least 95% of the organic agricultural ingredients. The main aim of new logo is to allow citizens to be more informed and then increase organic consumers' purchase with related positive economic effects on marginalized and rural areas. Hence, this paper has investigated the effects of paying attention to organic labelling on the intention to buy organic food products in Italy and how the intention drives to the final purchase. Moreover, it has also provided additional evidence regarding those psychological factors affecting the intention to purchase organic food products.

The results indicate that those consumers who pay attention to organic labeling when shopping are more likely to be willingness to buy organic food products. In addition, women and people with high incomes are those who are more likely to use organic labeling. And thus, they are to be willing to buy organic food products. On the other hand, the degree of paying attention to organic label is strongly and positively linked to the degree of knowledge. The results show that increasing organic knowledge is of paramount importance to increase the attention paid by consumers to organic label. As the Theory of Planned Behavior states, findings also show that other factors explain the intention to purchase organic food products such as attitudes towards their purchase. Consumers with positive attitudes towards the purchase are more likely to be willing to buy organic food. In particular, consumers who support buying organic foods are more likely to be willing to purchase them. However, in contrast with previous studies, the intention to purchase of Italian consumers is not affected by subjective norms or environmental attitudes when shopping. Finally, ethical dimension is considered one of main factors driving organic food purchases because Italian consumers who feel the ethical obligation to choose organic food products are the ones that finally purchase them. These results provide valuable information on consumers' purchase behavior of organic food products to help policy makers at European, National and Regional level. In particular, the results are very useful for policy makers when designing their respective organic farming policies for Italy. First, policy makers should take into account that household income still represents a barrier which limits the expansion of organic food products in the Italian market. Therefore, they should strongly support organic farmers by substantial subventions in order to decrease their costs of certifications and lead them to be more competitive in food market. Second, our study confirms the results of previous studies which showed that organic knowledge represents a key issue that organic farmers have to take into account in order to sell their products. To illustrate, organic food products should be supported by investments in promotion actions which will result in easy identifiable by consumers trough new organic labeling providing consumers more information of other organic food characteristics, such as the percentage of organic ingredients, animal welfare and GMOs free. Finally, since ethical aspects in organic food purchase influences consumers´ purchase decisions of organic food, future informational campaigns and promotion actions will also focus on making consumers more aware of the ethical dimension of organic food products. In particular, they also communicate the economic and environmental benefits on context where organic farmers operate in terms of the improvement of farmers' incomes and retaining population.

7. Acknowledgements

This paper has been written within of the project, funded by the European Union, "FOODLABELS_PIOF-GA-2009-253323

8. References

Aertsen, J., Verbeke, W., Mondelaers, K. & Van Huylenbroeck, G. (2009). Personal determinants of organic food consumption: a review. *British Food Journal,* 1140-1167, 111:10.

Ajzen, I. (1991). The theory of planned behaviour. *Organizational Behaviour and Human Decision Processes,* Vol. 50, pp. 179-211.

Arvola, A., Vassallo, M., Dean, M., Lampila, P., Saba, A., Lahteenmaki, L. & Shepherd, R. (2008). Predicting intentions to purchase organic food: The role of affective and moral attitudes in the Theory of Planned Behaviour. *Appetite,* Vol. 50 No. 2-3, pp. 443-454.

Bredahl, L. (2001). Determinants of consumer attitudes and purchase intentions with regard to genetically modified foods - Results of a cross-national survey. *Journal of Consumer Policy, 24,* 23-61.

Brunsø, K. Scholderer, J. & Grunert, K. G. (2004). Testing relationships between values and food-related lifestyle: results from two European countries. *Appetite,* 43, 195-205.

Canavari, M., Bazzani, G.M., Spadoni, R. and Regazzi, D. (2002). Food safety and organic fruit demand in Italy: a survey. *British Food Journal,* Vol. 104 Nos 3/4/5, pp. 220-32.

Chen, M.F. (2007). Consumer attitudes and purchase intentions in relation to organic foods in Taiwan: moderating effects of food-related personality traits. *Food Quality and Preference,* Vol. 18, pp. 1008-21.

Chen, M. F. (2008). An integrated research framework to understand consumer attitudes and purchase intentions toward genetically modified foods. *British Food Journal, 110 (6),* 559-579.

Chryssohoidis, G. M. & Krystallis, A. (2005). Organic consumers' personal values research: Testing and validating the list of values (LOV) scale and implementing a value-based segmentation task. *Food Quality and Preference,* Vol. 16 No. 7, pp. 585-599.

Dean, M., Raats, M. M. and Shepherd, R. (2008). Moral concerns and consumer choice of fresh and processed organic foods. *Journal of Applied Social Psychology,* Vol. 38 No. 8, pp.2088-2107.

de Magistris T. & Gracia A. (2009). The decision to buy organic food products in Southern Italy. *British Food Journal,* Vol. 110 Iss: 9, pp.929 - 947.

European Commission (1992) Regulation (EEC) No 2092/21, on organic production of agricultural products and indications referring thereto on agricultural products and foodstuffs

European Commission (2000) Commission Regulation (EC) No 331/2000 .Amending Annex V to Council Regulation (EEC) No 2092/91 on organic production of agricultural products and indications referring thereto on agricultural products and foodstuffs

European Council(2007) Regulation (EC) No. 834/2007 of 28 June 2007 on organic production and labelling of organic products and repealing Regulation (EEC) No 2092/91

European Commission (2010).Commission Regulation (EU) No 271/2010 on 24 march amending Regulation (EC) No 889/2008 laying down detailed rules for the implementation of Council Regulation (EC) No 834/2007, as regards the organic production logo of the European Union

Fotopoulos, C. & Krystallis, A. (2002). Purchasing Motives and Profile of the Greek Organic Consumer: A Countrywide Survey. *British Food Journal, 104(3/5):232-260*

Gil, J. M., Gracia, A., & Sanchez, M. (2000). Market segmentation and willingness to pay for organic products in Spain. *International Food and Agribusiness Management Review, 3*, 207-226.

Gracia, A. & de Magistris, T. (2008). The demand for organic foods in the South of Italy: A discrete choice model. *Food Policy*, vol. 33(5), pages 386-396, October

Grunert, K.G. & Brunsø, K. (1997). Food-related lifestyle: development of cross-culturally valid instrument for market surveillance", in Kahle, L.R. and Chiagouris, L. (Eds), *Value, Lifestyle and Psychographics*, Lawrence Erlbaum Associates, Hillsdale, NJ, pp. 337-54.

Guido, G., Prete, M., Peluso, A., Maloumby-Baka, R. & Buffa, C. (2010). The role of ethics and product personality in the intention to purcahse organic food products: a structural equation modeling approach. *International Review of Economics*, 57, 79-102.

Hajivassiliou, V. A., & McFadden, D. (1998). The method of simulated scores for the estimation of LDV models. *Econometrica, 66*, 863-896.

Hill, H. & Lynchehaun, F. (2002). Organic milk: Attitudes and consumption patterns. *British Food Journal*, 104 (7), pp. 526-542.

Honkanen, P., Verplanken, B., & Olsen S. O. (2006). Ethical values and motives driving organic food choice. *Journal of Consumer Behaviour*, 5 (5), 420-431.

ISTAT, 2008. I consumi delle famiglie. Webpage: https://www.istat.it

Lockie, G., Lyons, K., Lawrence, G. & Mummery, K. (2002). Eating 'green': Motivations behind organic food consumption in Australia. *European Society for Rural Sociology*, 42(1):23-40.

Magnusson, M. K., Arvola, A., Koivisto Hursti, U., Aberg, L. & Sjöden, P. O. (2003). Choice of organic foods is related to perceived consequences for human health and to environmentally friendly behaviour. *Appetite*, Vol. 40 No. 2, pp. 109-117.

Michaelidou, N., & Hassan, L.M. (2008). The role of health consciousness, food safety concern and ethical identity on attitudes and intentions towards organic food. *International Journal of Consumer Studies, 32*, 163-170.

Millock, K., Wier, M. & Andersen, L.M. (2004). Consumer's demand for organic foods-attitudes, value and purchasing behaviour, selected paper for presentation at the XIII Annual Conference of European Association of Environmental and Resource Economics, Budapest, 25-28 June.

Padel, S. & Foster, C. (2005). Exploring the gap between attitudes and behaviour. *British Food Journal*, Vol. 107 No. 8, pp. 606-25.

Poelman, A., Mojet, J., Lyon, D. & Sefa-Dedeh, S. (2008). The influence of information about organic production and fair trade on preference for and perception of pineapple. *Food Quality and Preference*, Vol. 19 No. 1, pp. 114-21.

Saba, A. & Messina, F. (2003). Attitudes towards organic food and risk/benefit perception associated with pesticides. *Food Quality and Preference*, Vol. 14, pp. 637-45.

Shaw, D. & Clarke, I. (1999). Belief formation in ethical consumer groups: an exploratory study. *Marketing Intelligence and Planning*, Vol. 17 No. 2 and 3, pp. 109-19.

Schifferstein, H.N.J. & Oude, P.A.M.(1998). Health-related determinants of organic food consumption in The Netherlands. *Food Quality and Preference*, Vol. 9 No. 3, pp. 119-33.

Smith, S. & Paladino, A. (2010). Eating clean & green? Investigating consumer motivations towards the purchase of organic food. *Australasian Journal of Marketing*, 18(2): pp 93-104.

Sparks, P. & Shepherd, R. (1992) Self-Identity and the Theory of Planned Behavior: Assesing the Role of Identification with "Green Consumerism. *Social Psychology Quarterly*. 388-399

Tarkiainen, A. & Sundqvist, S. (2005). Subjective norms, attitudes and intentions of Finnish consumers in buying organic food. *British Food Journal*, Vol. 107 No. 10-11, pp. 808-822.

Thøgersen, J. (2007a). Consumer decision-making with regard to organic food products. in Vaz, M. T. d. N., Vaz, P., Nijkamp, P. & Rastoin, J. L. (Eds.) *Traditional Food Production Facing Sustainability: A European Challenge; Ashgate.*

Thøgersen, J. (2007b) The motivational roots of norms for environmentally responsible behaviour. *Nordic Consumer Policy Research Conference. Helsinki.*

Thøgersen, J. & Olander, F. (2006). The dynamic interaction of personal norms and environment-friendly buying behavior: A panel study. *Journal of Applied Social Psychology*, Vol. 36 No. 7, pp. 1758-1780.

Thøgersen, J. (2002). Direct experience and the strength of the personal norm - Behavior relationship, *Psychology & Marketing*, Vol. 19 No. 10, pp. 881-893.

Tsakiridou, E., Mattas, K. & Tzimitra-Kaloglanni, I. (2006), The influence of consumer characteristics and attitudes on the demand for organic olive oil, *Journal of International Food & Agribusiness Marketing*, Vol. 18 No. 3-4, pp. 23-31.

Yiridoe, E.K., Bonti-Ankomah, S. & Martin, R.C. (2005). Comparison of consumer's perception towards organic versus conventionally produced foods: a review and update of the literature. *Renewable Agriculture and Food System*, Vol. 20 No. 4, pp. 193-205.

Verhoef, P. C. (2005), Explaining purchases of organic meat by Dutch consumers, *European Review of Agricultural Economics*, Vol. 32, pp. 245-267.

Von Alvesleben, R. (1997), *"Consumer behaviour"*, in Padberg, D.I., Ritson, C. and Albisu, L.M. (Eds), Agro-food Marketing, CAB International, New York, NY.

Zanoli, R. & Naspetti, S. (2002). Consumer motivations in the purchase of organic food: a means-end approach. *British Food Journal*, Vol. 104 No. 8, pp. 643-53.

Part 2

Systems and Farmers

7

Contesting 'Sustainable Intensification' in the UK: The Emerging Organic Discourse

Matthew Reed
Countryside and Community Research Institute, The University of the West of England
UK

1. Introduction

Over the past 15 years organic food and farming has been a consistent topic of study across a range of disciplines; geography, sociology, social psychology and marketing as well as the agricultural and ecological sciences. Some of these interventions were based on the hope that the organic approach to agriculture could improve not only the farmed environment but also social relations between producers and consumers. From this point of optimism there came an almost instant backlash that moved from the critical to the condemnatory. For several years this was in contrast to the ever-rising commercial fortunes of certified organic products that were becoming increasingly favoured by consumers. The recent economic crisis appears to have stalled the commercial success of organics, leaving it out of fashion with consumers, as well as with many scholars. Global critiques such as those of Holt Giménez and Shattuck, leave organics in an ambiguous position, being labelled part of the neo-liberal project despoiling people and the planet alike, yet badged as 'agro-ecology' being part of what they define as the most radical attempts to redefine the global food system (Holt Giménez & Shattuck 2011). Global bodies such as the UN and World Bank can find that the 'food crisis' of our times might be resolved by forms of sustainable agriculture (De Schutter 2011, IAASTD 2009). Yet, many activists can find that organic has become so degraded as to be simply the choice between one form of corporate branding and another (Patel 2007).

The purpose of this paper is to argue that the organic movement is entering into a new phase of activity and that to understand these changes scholarly accounts need to be cognisant of it as a social movement and sensitive to socio-spatial differences within the movement (Reed 2010). By considering how the movement's organisations have discussed and contested attempts to intensify British agriculture this paper examines new permutations in the discourse of organic agriculture. This new permutation has developed in opposition to, and in tension with, the drive to 'sustainably intensify' British agriculture. Broadly the British state, with support from networks of scientists and corporations, has been attempting to frame the future of the food system as being reliant on a raft of new technologies, which are being resisted by a range of NGOs and social movements. As Holt Giménez and Shattuck have argued these northern, middle class movements - such as the organic movement - often straddle reformist and progressive positions regarding the food system, therefore present opportunities for change that might be of wider significance.

Therefore arguments formed and forged in one context may become relevant for the food system more widely. Before addressing that discussion this paper considers the academic critiques of organic food and farming, as well as the web of relationships within which the movement operates. The paper concludes by considering how the emerging configuration of the movement may start to reshape the socio-spatial relations of food production and consumption.

"*The crisis in 2005–8 was not a blip, but creeping normality* (Lang 2010:95)".

As Tim Lang has persuasively argued, many of the features of the current crisis of food are not novel but extensions of the trajectory of the food system in the twentieth century (Lang 2010). Yet there are developments, some of which are the outcomes of that trajectory that mean the context within which these changes are taking place needs to be understood (Castells 1997,1998). Lang emphasises the increasing environmental pressures that are potentially going to limit the volumes of food produced. Yet, as will be discussed below, his style of critique shares elements that are common with those who would generally oppose his positions. By using the analytical tools derived from the social sciences we can perhaps mark those trends and interventions that are of lasting significance and those that the transitory. The challenge is to do so in a way that does not use meta-theory to produce narratives that mask differences and moments of resistance. Under the rubric of 'neo-liberalism' and those trends in 'resistance' to it, are flows and possibilities that suggest a more nuanced picture.

2. Social movements

Social movements like the social networks that they constitute and are created by, are conceptually mid-level phenomena (Crossley 2002, Melucci 1996). For several influential theorists they are important agents of change, Castells argues that they are 'symptoms of who we are' in the Information Age. Whilst Charles Tilly argued that they mark a particular form of the contentious politics and the presence democratic opportunities (Castells 1997, Castells 1998, Tilly 2004). Most germane to this paper is the role that social movements play in theories of the contemporary food system, and hence global capitalism, as analysed in the food regime approach (McMichael 2009a, McMichael 2009b). Food regime theory accounts attempt to analyse the stabilities that constitute originally the political economy but increasingly the political ecology of a globalised food system that is based on inequalities both within and between nations and communities (Campbell 2009). Through periodising the operations of the trade in food products as well as the consequences of different forms of agricultural production, processing and distribution, food regime theory seeks to understand the tensions and paradoxes within a regime of accumulation (Burch & Lawrence 2009). In doing so it aims to locate sites and agents of resistance to, as well as mechanisms of, the exploitation of people, animals and eco-systems.

Social movements act in regime theory to either validate or challenge the food cultures of particular food regimes. They can act at moments in the transitions between food regimes either against, or in concert with, other powerful actors. Friedmann in her accounts identified durable food products transported over long distances as central to the food regime of the 1990s, so she saw movements taking 'local' and 'seasonal' as offering a possible locus of resistance. Although she argued that these could also be appropriated in a regime based around corporate dominance, as observed by Guthman (Guthman 2004a). McMichael more recently has identified the globalised peasant and small farmers' movement Via Campesina as a movement contesting the neo-liberal conceptualisation of food as a commodity with their arguments for food to be re-localised through arguments for 'food sovereignty'. Whilst this argument prioritises those who are exploited and marginalised in the global South, considerable opportunity is identified for these movements to act in co-ordination with Northern social movements.

3. The organic food and farming movement

With its origins in the colonial encounter of Western agriculture with its Asian counterpart, and a separate hermeneutic tradition from German speaking Europe, organic agriculture started in the 1920s (Conford 2001, Reed 2010, Vogt 2007). By the 1930s it had networks of discussion and a few experimental farms across Germany and the British Empire. It was only in the post-war period that organic farming began to spread more widely, partly as a response to the green revolution and partly through the emerging organic movement. Configured, as I have argued elsewhere, as a cultural movement the organic movement lacked for many years the confrontational tactics of many other social movements but worked on exemplars of alternative agricultural practice and increasingly on providing organic products (Reed 2010). During the late 1960s and the early 1970s the Soil Association, the main organisation of the British organic movement, saw a range of radical environmental thinkers and activists clustered around it, ranging from the eco-socialist Barry Commoner, through the conservative Edward Goldsmith by way of its President E F Schumacher (Reed 2004, Reed 2010).

The adoption of EU organic production standards saw these goods move from health food stores and farm shops towards the major retailers, with subsequent rapid growth for those elements of the movement involved in farming (Buck *et al.* 1997). At this point the organic movement found itself at the forefront of a direct contest with its opponents for the first time, as it became a mover within, and tribune for, the protests against Genetically Modified/Engineered plants. These protests demonstrated the global spread of the organic movement until this point and locked it into what Campbell has described as a 'binarism' with GM agriculture (Campbell 2004).

Just as these protests saw these technologies largely removed from Europe and fiercely contested elsewhere, many scholars and activists were positing that organic farming no longer offered any resistance to the dominant forms of agriculture (Guthman 2004b, Lilliston & Cummins 1998, Rigby & Young 2001). Tovey had argued that in the Irish example the appearance of production standards had seen the institutionalisation of the movement and this was confirmed by Moore who demonstrated that many organic growers were moving to a 'post-organic' status to find new cultural space (Moore 2005, Tovey 1997). In part these differences can be explained by the local trajectories of different national organic movements. Although the protest actions in the UK that provided elements of a repertoire of protest that was widely emulated, suggesting divergent flows within and between national movements (della Porta & Tarrow 2005, Reed 2010). Guthman's prescription for subsidies for organic production, stronger regulations and more technical support appear very similar to initiatives common in Europe, yet the message of the 'conventionalisation thesis' has been broadly applied without these caveats (Formartz 2006, Patel 2007, Pollan 2006). There has also been much sport in what Johnston and Szabo have described as "scholarly cynicism about affluent food consumers and their selfish motivations" (Johnston & Szabo 2010:14)

4. Sustainable Intensification

"We head into a perfect storm in 2030, because all of these things are operating on the same time frame, ...If we don't address this, we can expect major destabilisation, an increase in rioting and potentially significant problems with international migration, as people move out to avoid food and water shortages", (Sample 2009)

The present food crisis, which started in early 2008, was triggered by rapidly rising international prices of grains, propelled by a series of short-term factors forming a "perfect storm"; more importantly, however, many underlying longer-term factors had been brewing in the market for some time, making the crisis inevitable. (United Nations 2009:26)

What for the United Nations was an inevitable crisis, is for Sir John Beddington, Chief Scientist to the UK government, a harbinger of an even more perfect storm of globalised disorder and hardship as food begins to run short. A new consensus has been rapidly appearing within elite groups, that food supplies are going to be compressed and this is likely to become a prominent feature of the next decades. The analysis of the flows and forces that led to the vortex of this storm forming quickly became divided between those who view it as the product of the pressures stemming from the success of development and resulting environmental pressures and those who view it as a product of the globalised market in food. The former group tend to emphasise the importance of technological innovation, underpinned by applied scientific research to increase the productivity of agriculture. For them the pressures of inexorable global population growth to the peak of 9 billion in 2050, in tandem with the environmental pressures of global warming means that the challenge is beyond distribution but of the absolute physical lack of agricultural products. Often self-consciously they are echoing the arguments that launched the green revolution, arguing for a renewal of that project but with a greater attentiveness to the environmental consequences of such intensification of production.

The arguments that are most closely associated with the discourse of food security poses three questions, that of the *access* to food, its overall *availability* and its relative *affordability*. Within this discourse, questions of the demand for food and the conformation of those foodstuffs, the power of the major market players and global management of those resources are reified. It also tends towards the Malthusian, in that population dynamics are almost always negative in their consequence, in that high numbers are an unmitigatedly bad outcome and an aging population is just as problematic. In this we can see the arguments around food security as a form of environmentalism, conforming closely to what Dryzek has previously classified as 'administrative rationalism'. In this discourse liberal capitalism and the administrative state are reified, with nature subsumed to human problem solving and the key agents of change are experts and/or managers motivated by the public interest, the 'public' being a unitary group rather than a range of constituencies (Dryzek 1997).

Those who see the crisis as the result of the operations of the global market target a range of actors and processes. Walden Bello, looking at the crisis from the perspective of the global south points to the extension of liberal capitalism through the structural adjustment programmes of the 1980s and 1990s that brought local food producers into the global market and broke down the infrastructures looking to develop national capacity (Bello 2009). Others point directly to the role of speculators in causing the volatility, as investors have poured into complex speculative tools (Kaufman 2011). Yet, these critiques are often unable to adopt positions of diametric opposition as they share some of the premises of the arguments of their opponents. Most share the opinion that the planet is approaching its natural limits and many hold even more pessimistic arguments about global warming and the vicissitudes of global warming (Holden 2007, Pfeiffer 2006).

For many hailing the need for a new or doubly green revolution, the appeal of this sort of administrative rationalism is apparent, and to a degree those groups thet opposed, which opposed the aspects of the green revolution, are opposed to this renaissance. Yet, it is apparent that both groups share many of the same epistemic assumptions; the finite limits of the planet, the demographic pressures and the impending peril of climate change. Equally aspects of their lexicon are shared, dominated by conservation of resources, shades of green and the importance of biological processes. It is over questions of participation, the forms of technology to be deployed, the role of liberal markets and national autonomy that they diverge.

5. Collapsed in the aisles

In fact, much of the 'organic' produce shipped in from around the world and across the UK today carries no sense of connection with its geography or its farmers. It is as anonymous as the majority of conventional chemically produced foods, as dull in flavour and as lacking in nutritional vitality (Rose 2010)

Sir Julian Rose used his position as a pioneer, having farmed organically since 1975, and an article in *The Ecologist* magazine to point out the failings of the contemporary organic industry. His answer was for the movement to return to its roots and to stop chasing a "big branded chimera". In this Sir Julian echoed academic and activists critiques of organic food. The sociologist Raj Patel has condemned the difference between organic and non-organic food as the choice between 'Pepsi and coke' (Patel 2007). Whilst Heath and Potter condemn organic products because organics they argue is based on unfounded health claims and the difference in price purchases only social distinction, in contrast to their quixotic example of a hybrid car. Although Patel, Heath and Potter are drawing on the North American experience for their critique Rose's follows the same pattern although confining itself the UK. Organic food has become a commodity like any other by being sold in supermarkets, and for Heath and Potter they are worse because they claim a spurious moral status and so create a socially destructive cachet or 'cool'. Their arguments are not based on evidence but rather an argument that the system of retailing is, as Rose argues, "Orwellian".

Attacks such as Rose's are hardly new in *The Ecologist* but the difference was the context of this criticism, as sales of organic food in the supermarkets and beyond were falling. Organic sales began to fall as soon as the recession began with sales falling by 13.9% at the of 2009, after rising by 1.7% overall in 2008 and showing a signs of a return in 2010 as month by month comparisons moved from -12% to -8%. This was not a uniform decline, with babyfood and milk continuing to increase sales throughout the period, whilst organic prepared foods, meat and brands such as Duchy Originals, owned by Prince Charles, being particularly hit (The Soil Association 2010). Although the opponents of organics in the media and farming industry sniped, analysts in the retail sector remained confident in the resilience of the organic sector.

Shoppers have not performed a u-turn on ethics, so the challenge for organic is to make sure that communication of its benefits is clear and consistent. If they get that right, it would be sensible to assume that volume sales could pick up as the economy recovers. (Grocer 2009)

The decline in sales certainly caused difficulties to businesses that were planning for continued expansion and particularly for meat producers, with the most high profile victim being Price Charles who found his brand being rescued by the supermarket 'Waitrose'. The amount of land organically managed rose in 2009, to 4.3% of the UK's farmland. The geographic distribution of organic farmland remains complex with a strong increase in Wales reflecting that nations agricultural policies and a continuing strong presence in the South West of England.

The plunge in sales had a galvanising affect on the organic industry in the UK, as it had previously tended to allow the campaigning groups to promote organic whilst individual businesses focused on marketing their own brands (Reed 2009). This changed with the formation of the organic trade board (OTB)[1], which as well as seeking to represent the industry, looks to share market research, promote effective communication with consumers and to improve the evidence base for organic products. As part of this the OTB along with the environmental charity Sustain promoted the OrganicUK campaign to raise funds that would be matched by the EU to promote organic products generically in the UK, with the announcement of a 3 year promotional campaign costing £2million in July 2010. In the autumn of 2010 this collective effort was initially eclipsed by one of the donors to OrganicUK solo effort. The Yeo Valley dairy used short advertising slots during the popular TV talent show the 'X-factor' to trail an on-line video of some of its farmers rapping. In the first two days it had secured over 350,000 on-line views which had risen to 1.4 million by mid-December on the dairy's own YouTube channel, inevitably - Yeotube. Its products during this period offered the chance to win tickets to the X Factor, as Yeo Valley spent £5million attracting a youthful audience for organic milk[2].

The recession in organic sales saw the UK organic industry organise itself and move into promotional activities in many ways clarifying the role of the charities such as the Soil Association that had previously conducted much of this work. That it was the diary businesses, the least effected by the recession and the largest enterprises, that were at the forefront of these developments suggests something of the future direction of the organic industry. Until the recession much of the advertising of organic products had been marked by elitism, with branding aimed at more affluent consumers (Cook, Reed et al. 2008). For many in marketing and retailing this dovetailed with the higher costs of production in some organic systems, resulting in organics to be positioned within stores and brandscapes at the more expensive end of product ranges. Some, such as Riverford's Guy Watson consistently argued against this approach and the damaging impacts it had on the organic market, but until the recession their warnings went unheeded. The first advertising campaign resulting from this initiative 'Why I love Organic', changed the tenor of previous organic marketing by deliberately featuring working men, alongside celebrity endorsements and social media links. Without the upward pressure of rising incomes, the previous marketing strategy was exposed. Re-orientated by actors such as the OTB the emerging strategy is less elitist as it aims at penetrating the mass-markets often disparaged by activists.

6. Super dairies

In 2009 most of England's dairy cattle lived on farms that ran herds of between 100-200 animals, with the second largest group were those in herds of over 200 animals. Against

[1] www.organictradeboard.co.uk

[2] http://www.yeovalleyorganic.co.uk/

the backdrop of an overall decline in the number of dairy animals through the decade, the role of larger herds in dairy production had been growing[3]. Many of the herds recorded as being over 200 animals, are kept on separate farms but owned and managed by one enterprise. The proposal for a single dairy unit of over 8100 cows, managed as one unit, a step of 5800 more cows than the next largest unit, signalled a major leap in the scale of farms producing milk in the UK. Those behind the proposal argued that they would be able to realise economies of scale, in that the unit would be generate energy from anaerobic digestion facilities on site and transportation costs would be lowered, with the welfare of the animals welfare of the animals maximised maximised by being kept mainly indoors, with only limited summer grazing[4]. The farmers behind this proposal were open that their inspiration was the similar dairy units that they had seen in Wisconsin, in the United States. In December 2009 an application for the requisite planning permission was lodged with North Kesteven District Council.

The Nocton proposal tripped across the wires of numerous groups and cultural boundaries that were not always found in common cause. Much of the debate was defined within the cultural terms of 'Britishness' - that the UK had a distinct tradition of dairy farming and that this represents a good example to other countries, post-imperial agricultural leadership. The World Society for the Protection of Animals (WSPA), a London based umbrella body for a coalition of animal protection societies, launched a campaign against the use of milk from battery farm cows. The "Not in my Cuppa" used of the role milk in the national beverage - the cup of tea - as the fulcrum for arguments about the impacts of animal welfare for cows in such a system as proposed at Nocton. In this campaign they were joined by the Compassion in World Farming, Friends of the Earth, the Campaign to Protect Rural England, The Soil Association, 38 Degrees and a group from the area near the proposed unit - Campaign Against Factory Farming Operations. These latter groups widened the arguments to the future of farming, the impact on the environment - locally and globally - as well as the conservation of the traditional English landscape, with some proposing the positive solution being the adoption of organic milk and dairying.

It was the statutory body charged with protecting the environment, the Environment Agency, which withheld its permission over concerns about the amount of manure being generated on the farm and its likely impact on the local watercourses. In April 2010, the application was withdrawn in the light of this advice and in November a revised application for a unit of 3,770 cows was submitted, only to be withdrawn in February 2011. In a statement from Nocton's developers, they cited the lack of research that they could draw on to persuade the Environment Agency that the farm did not represent a threat to the local aquifer. They were at pains to point to their relationship with the Agency:

We believe the Environment Agency has not acted under any pressure in reaching this decision and that no undue influence from other individuals or organisations has been brought to bear; any claims to this effect would be both disingenuous and self-serving (Nocton Dairies 2011).

The district council, as the ultimate planning authority, made it clear that it had concerns about the housing of workers, the loss of amenity to local people and the wastes from the

[3] http://www.dairyco.org.uk/datum/on-farm-data/cow-numbers/uk-cow-numbers.aspx

[4] http://www.noctondairies.co.uk/

unit. Those behind the proposal knew that the opinions of the Environment Agency were central, as they were unlikely to be overturned at appeal, whilst those of the district council could be.

This application was about far more than the enterprise alone, as the statement withdrawing the application made clear:

The challenge has been laid down to the farming industry to produce more with less. We need leadership to help us do this and proactive advice from regulatory experts – only a practical, informed and 'can-do' approach will move this whole agenda forward (Nocton Dairies 2011).

Their opponents were also clear that victory for Nocton would have been a 'tipping point' and the end of "Our smaller-scale, predominantly pasture-based dairy farmers, under whose stewardship Britain's dairy cows have grazed countryside pastures for generations" (Morris 2011). Of equal significance was that the British public had rejected intensive animal husbandry and "Britain is a world beacon for farm animal welfare", so a failure would encourage others to adopt intensive technologies.

The Soil Association and the WPSA published a report in April 2011 pointing out how large-scale developments such as Nocton and similar pig unit at Foston in Derbyshire would put many smaller farmers out of business. Foston, a proposal for a pig unit of would breed around 25,000 young pigs a year and was the subject of the 'Not in my Banger' campaign, a clone of the 'Not in my Cuppa' one targeting Nocton. The Soil Association reported argued that as the domestic supply of milk was already fulfilled by, Nocton would have to displace existing producers by undercutting them on price. Using industry figures, they argued 60-100 average sized farm businesses would be displaced by Nocton's entry into the milk industry. The Soil Association had been campaigning against intensive animal production in the UK since the 1960s, but the arguments against Nocton, and Foston, represent a new permutation in their discourse. Previous arguments have been concerned with the technologies of confined production; the new permutation brings into play the scale of this deployment. The Soil Association is beginning to defend explicitly family farming:

These smaller dairies and pig producers will be ideally suited to serving local markets, and will often represent a family's main or at least an important source of their income. The families running many of these farms will have been producing milk and pork for generations (The Soil Association 2011:4).

7. Cloned

As the UK got used to the idea of a coalition government it was revealed in August 2010 that a number of cloned cattle had entered the food chain. A cattle breeder in the Scottish Highlands had privately imported cloned embryos to augment his herd's line and at least one of these animals had found their way into the food chain. Given the UK's history of cattle related food and health crises this was met with newspaper headlines and an investigation by the Food Standards Agency (FSA). As a novel food product, the safety of which had yet to be assessed it should not have entered the food chain, although the consequences for non-compliance were opaque. Coincidentally the EU parliament had voted in July for a moratorium on cloned animals or their progeny or products entering the food chain, although the Commission did not share that position. The topic gained pace as in late November, after a review initiated by the FSA, the Advisory Committee on Novel Foods and Processes (ACNFP), reported that the produce from and cloned animals themselves were 'unlikely to present a food safety risk'. It also appealed for more evidence to be able to present findings with greater certainty and that consumers might want a labeling scheme in place. This opened the way for the FSA board to discuss the matter, and in turn make a recommendation to the Minister, which it did in May 2011, that the progeny of cloned animals be allowed in the food chain. Even if the Minister approved clones and their produce as 'safe', many anticipated considerable problems with public acceptance of these products, although press reports suggest that the ministers were not minded to press for labeling.

Despite the change in government this move represents continuity with the previous administration's determination to have the administrative and legal framework in place for genetically engineered or modified plants despite no domestic market or demand to grow such crops. The FSA initiated the safety review as no farmer or business had done so, and set the procedure in motion to have clones found to be safe. Even though research commissioned by the FSA had found widespread opposition from the public to cloning, and a belief that the system of approval was not adequate:

- There is a major mismatch between the methods used by regulatory
- authorities to assess food safety and the public's perception of what
- is needed. Participants wanted to see methods for assessing food
- safety that were analogous to the approach used in clinical drugs trials (Creative Research 2008 :2)

In this the research echoed that of the more formal and larger consultation about plants 'GM Nation' that there was little interest in such crops being planted. As with GM plants none of the supermarkets were prepared to endorse the use of clones and had previously made unambiguous statements about avoiding clones or their products. The stance of the government appeared to be to leave the opportunity in place to take up GM plants, and more recently clones, in anticipation of domestic demand for such products. This is fully in accordance with the discourse of food security discussed above, where technologies managed by experts in the public interest will address the upcoming crisis.

8. Discussion

In their categorisation of the responses to the global food system Holt Gimenez and Shuttack note that attention needs to be paid to the specific circumstances of movements and the opportunities for alliance. This paper has aimed to do just that and then suggest how these might have a wider influence, as they are diffused through the global organic movement. On occasions proponents have argued because the organic movement in their locality or jurisdiction has displayed particular tendencies then all organic movements across the planet will follow suit. It may be that organics in North America, at a federal level, or within a particular certification system has become dominated by corporate interests. Similarly individual organic farmers are part of the most radical of groups and champions of the broadest change to the agricultural system - such as Jose Bove. The British organic movement has displayed tendencies that suggest an accommodation with the food system; over 70% of organic products in the UK are sold through supermarkets. Yet, it has also displayed the more strident opposition to GM technologies, continues to contest other technologies that seek to intensify British agriculture and battle the corporate domination of agriculture more broadly.

Joining with other movements, across ideological and spatial divides remains a challenge for the British movement. Although much of the English language critique of the food system stems from North America, on the ground the differences between European food system and that of the US robs these criticisms of practical application. The broad ideology might be shared, the rhetoric and imagery appeal, but the gap in practice and policy is too wide to have much practical bearing. As the debate about air freighting organic produce demonstrates, the trade-offs between environmental benefits and social goods are difficult; with the Soil Association ultimately preferring to demonstrate solidarity rather than environmental purity. Although influential individual enterprises such as the Riverford family of box schemes have taken a different route, choosing not to airfreight (Watson and Baxter 2008). At the same time global trends, such as the embedded water in meat and dairy products are less germane when considering the wet, temperate uplands of the west of England and Wales. Here often the most sustainable form of agriculture is extensive grass fed animal husbandry. Weaving a sustainable food system will be in part attentiveness of the specifics of place and culture, but solidarity across distance will also be important.

The focus of the organic movement also remains locked onto the food system of the twentieth century, with questions of agriculture production trumping those of processing, distribution and consumption although these latter concerns are moving up the agenda. Although occasionally the role of poorly paid, migrant and abused labour has been raised in relation to the food industry; this has not been taken up the domestic organic movement. The IFOAM review of organic standards may be put social justice into the core of organic aspirations, it has yet to find its way into certification standards. Repenting from its up-market image during the boom years at the turn of the century, the newly organised organic sector is determined to be more egalitarian and popular, social claims about organics have yet to find its way into the promotion of organic products. It also needs to construct new roles for activists, producing and consuming organic products are quietist roles - the farmers literally tying people to the farm and the latter at most a supporting role. Experiments in mass share holding of a farm, or more direct forms of protest have proved to be popular within the movement suggesting that new roles could be quickly filled if more widely articulated.

9. Conclusion

By insisting on class positions tied to food movements, Holt Gimeniz and Shuttack remind us that whilst northern consumers are relatively powerful actors in the food system, this power is circumscribed by political opportunity and the greater powers of corporate actors. Hence protests and mobilisation tend to be reactive, contestations of the actions of others rather than initiatives from the movement. The protests and lobbying against the introduction of clones and mega-dairies continues a long history of fighting the development of mainstream agriculture - innovation by innovation. One area where the movement has been able to make considerable strides in the development of knowledge and interventions supportive of sustainable agriculture. Although the global movement has been littered with research farms, test plots and applied research, it is only in the past twenty years that a sustained effort has been put in place to develop peer-reviewed scientific knowledge about organics. This represents in part a retreat by the movement away from an insistence on 'wholistic' enquiry into organic farming, that made investigations both complex and often outside the parameters of the existing journal system. It also represents the determination of scholars allied with the movement to provide the movement with not only the practical knowledge to farm organically but also to argue for organic in policy circles. After nearly 80 years of work the organic movement is increasingly able to prove its case, without presuming that decisions are always made with regard to evidence.

The British organic movement is likely to remain straddling the reformist and progressive tendencies within the food system, until the political opportunity structure within the food system opens. If the history of the movement is a guide then this is not solely about the dominant food system but also crises within the movement itself, the disintegration of the late 1960s saw the emergence of organic standards; the decline in the early 1990s saw the introduction of box schemes and the greatest headway was made in opposing the introduction of GM crops. The British organic movement continues to display the potential to be influential actor in reforming the food system (Reed 2010). The form and timing of its next significant intervention is not apparent but its continued activity suggests that it respond will to the next significant opportunity.

In the past decades the British organic movement has provided the global organic movement and those movements allied to it with a number of examples that have created opportunities. The most significant has been to pioneer the construction of a market based on a certification scheme controlled by a social movement. This development created the space and resources both physical and ideological for the wider movement to grow rapidly. The forms of protest developed by British protestors against GM crops, which had in turn been adapted from Australian tactics, were widely emulated as the dispute was diffused globally. This suggests that whilst northern movements may play a particular role in the global social movements that not all national movements are equally influential or positioned to be so. Whether it is the legacy of the empire or that British English is a variety of the *lingua franca* of the dominant global language, the UK's organic movement appears to have historically enjoyed a particular place of influence. This suggests that innovations within and by the British movement may have a wider importance for the global movement, making it worthy of continued study and the investment of energy by activists looking to make a change to the global food system.

10. References

Bello, W. (2009) *The Food Wars* Verso, London, ISBN, 13:978-1-1-884673315

Buck, Getz and J. Guthman (1997) From Farm To Table: The Organic Vegetable Commodity Chain Of Northern California. Sociologia Ruralis, 37, 1, ISSN 0038-0199, 3-19

Burch, D. and G. Lawrence (2009) Towards A Third Food Regime: Behind The Transformation. *Agriculture and Human Values*, 26, 4, ISSN 1572-8366, 267-279

Campbell, H. (2004) Organics ascendant: Curious resistance to GM. in R. Hindmarsh and G. Lawrence eds., *Recoding nature: Critical perspectives on genetic engineering*, University of New South Wales Press, ISBN 0868407410, Sydney, Australia.

Campbell, H. (2009) Breaking new ground in food regime theory: Corporate environmentalism, ecological feedbacks and the 'food from somewhere' regime? *Agriculture and Human Values*, 26,4, ISSN 1572-8366, 309-319

Carson, R. (1991) *Silent Spring*, Penguin, ISBN 978-0140138917, London

Castells, M. (1997) *The Power Of Identity*, Blackwells, ISBN 978-140519687, London.

Castells, M. (1998) *End of the Millennium*, Blackwells, ISBN 978-0631221395, London.

Collier, P. (2008) The Politics Of Hunger. *Foreign Affairs*, 87,6, 67-79

Conford, P. (2001) *The Origins Of The Organic Movement*, Floris Books, ISBN 978-0863158032, Edinburgh.

Cook, G., M. Reed, et al. (2008). ""But it's all true!" Commercialism And Commitment In The Discourse Of Organic Food Promotion." *Text and Talk* 29,2, ISSN 1860-7349, 151-173.

Creative Research (2008) *Animal cloning and implications for the food chain findings of research among the general public.* Pp. 137 , Food Standards Agency, London

Crossley, N. (2002) *Making Sense Of Social Movements*, The Open University Press, ISBN 978-0335206025, Milton Keynes, Bedfordshire.

Dryzek, J. (1997) *The Politics of the Earth*, Oxford University Press.

De Schutter, O. (2011) *Report Submitted By The Special Rapporteur On The Right To Food, Olivier De Schutter*. Pp. 21 in, UN General Assembly Human Rights Council, New York

della Porta, D. and S. Tarrow (2005) Transnational processes and social activism: An introduction. Pp. 1-17 in D. della Porta and S. Tarrow eds., *Transnational protest and global activism*, Rowman and Littlefield, ISBN 0-7425-3587-8, London.

Formartz, S. (2006) *Organic Inc. Natural Foods And How They Grew*, Harcourt, ISBN 978-0156032421, New York.

Grocer, T. (2009) Organic: Nature's way? in, *The Grocer*

Guthman, J. (2004a) Agrarian Dreams. *The Paradox Of Organic Farming In California*, University of California Press, ISBN 978-0520240957, Berkley, California.

Guthman, J. (2004b) The Trouble With 'Organic Lite' In California: A Rejoinder To The 'Conventionalisation' Debate. *Sociologia Ruralis*, 44, 3, SSN 0038−0199, 301-316

Holden, P. (2007) *One Planet Agriculture*. The Soil Association, Bristol, England

Holt Giménez, E. and A. Shattuck (2011) Food crises, food regimes and food movements: Rumblings of reform or tides of transformation? *Journal of Peasant Studies*, 38, 1, ISSN 0306-6150, 109 - 144

IAASTD (2009) *Synthesis Report. A Synthesis Of The Global And Sub-Global IAASTD Reports*. B. D. McIntyre, H. R. Herren, J. Wakhungu and R. T. Watson eds., International Assessment of Agricultural Knowledge, Science and Technology for Development, ISBN 978-1597265393.

Johnston, J. and M. Szabo (2010) Reflexivity And The Whole Foods Market Consumer: The Lived Experience Of Shopping For Change. *Agriculture and Human Values*, ISSSN 1572-8366, 1-17

Kaufman, F. (2011) How Goldman Sachs Created The Food Crisis. Foreign Policy,

Lilliston and R. Cummins (1998) Organic vs 'organic': The corruption of a label. *The Ecologist*, 28, 4, 195-200.

McMichael, P. (2009a) A food regime analysis of the 'world food crisis'. *Agriculture and Human Values*, 26,4, ISSN 1572-8366, 281-295

McMichael, P. (2009b) A food regime genealogy. *The Journal of Peasant Studies*, 36, ISSN 0306-6150, 139-169

Melucci, A. (1996) *Challenging Codes. Collective Action In The Information Age*, Routledge, ISBN 978-0521578431, London.

Moore, O. (2005) What Farmers' Markets Say About The Perpetually Post-Organic Movement In Ireland. in G. Holt and M. Reed eds., *Sociological perspectives of organic agriculture: From pioneers to policies* CAB International, ISBN 978-1845930387, Wallingford, Oxfordshire, UK

Morris, S. (2011) Why the victory over the nocton super-dairy is only the beginning. in, *The Ecologist*

Nocton Dairies (2011) "Ambitious plans for UK's largest dairy farm withdrawn."

Patel, R. (2007) *Stuffed And Starved. Markets, Power And The Hidden Battle For The World's Food System*, Portobello, ISBN 978-8172237615, London.

Pfeiffer, D.A. (2006) Eating Fossil Fuels: Oil, Food And The Coming Crisis In Agriculture, New Society Publishers, ISBN 978-0865715653, Canada.

Pollan, M. (2006) *The Omnivore's Dilemma*, Penguin, ISBN 978-0747586838, London.

Reed, M. (2004) *Rebels for the soil: The lonely furrow of the soil association 1943-2000*. PhD, University of the West of England

Reed, M. (2009) For whom?: The governance of the british organic movement. *Food Policy*, 34, 2, ISSN 0306-9192, 280-286

Reed, M. (2010) *Rebels For The Soil - The Rise Of The Global Organic Movement*, Earthscan, ISBN 9781844075973, London.

Rigby, D. and T. Young (2001) The development of and prospects for organic farming in the UK. *Food Policy*, 26,6, ISSN xxxx, 599-613

Rose, J. (2010) Organic farming has sold out and lost its way. in, *The Ecologist*,

Sample, I. (2009) World Faces 'Perfect Storm' Of Problems By 2030, Chief Scientist To Warn. in, *The Guardian* (London)

The Soil Association (2010) *Telling Porkies. The Big Fat Lie About Doubling Food Production*. The Soil Association, Bristol.

The Soil Association (2010). *Organic Market Report 2010*. The Soil Association, Bristol.

The Soil Association (2011) *Old MacDonald Had A Farm. The Possible Impact Of Proposed Mega Dairies And Massive Pig Factories On The Small Family Farm*. The Soil Association, Bristol

Tilly, C. (2004) *Social Movements 1768-2004*, Paradigm Publishers, ISBN 978-1594510434, London.

Tovey, H. (1997) Food, Environmentalism And Rural Sociology: On The Organic Farming Movement In Ireland. *Sociologia Ruralis*, 37, 1, ISSN 0038-0199, xxxx.

United Nations (2009) *World economic situation and prospects 2009 – global outlook 2009* –. in, (New York: United Nations)

Vogt, G. (2007) The origins of organic farming. Pp. 9-30 in W. Lockeretz ed., *Organic farming: An international history*, CABI, ISBN 978-0851998336, Wallingford, Oxfordshire, UK.

Watson, G. and J. Baxter (2008). *The Riverford Farm Cook Book.* Fourth Estate London.

8

Farmers' Attitudes Towards Organic and Conventional Agriculture: A Behavioural Perspective

David Kings[1] and Brian Ilbery[2]
[1]*The Abbey, Warwick Road, Warwickshire*
[2]*Brian Ilbery, University of Gloucestershire, Oxstalls Campus, Oxstalls Lane, Longlevens, Gloucester*
UK

1. Introduction

Using a modified behavioural approach, this chapter examines organic and conventional farmers' relationship with the concept of food security. The World Food Summit (1996) defined food security as existing: 'when all people at all times have access to sufficient, safe, nutritious food to maintain a healthy and active life'. Additionally, the concept is commonly thought of as including both physical and economic access to food that meets people's dietary needs as well as their food preferences. In recent decades, food security has been usually associated with developing countries (Frow et al., 2009). This chapter, however, is primarily concerned with aspects of food security in the UK and thus a European model which expects farmers to provide other societal benefits such as biodiversity, environmental protection and food safety. Such a model aims to satisfy consumers' demand for 'healthier and more flavoursome food of higher nutritional value, produced by more environmentally friendly methods' (Brunori & Guarino, 2010).

This chapter is based on the proposition that the attitudes and behaviours of organic farmers may differ from those of conventional farmers, especially in relation to farming, the environment and food security. A second proposition is that farming systems towards the organic end of the agricultural spectrum may appeal first and most strongly to farmers already attuned to environmental ideas. The chapter aims to compare the perceptions, attitudes and behaviours of those farmers loosely labelled 'organic' and 'conventional' in central southern England, especially in relation to their attitudes and values towards farming, the environment and food security. More specifically, the research has the following supporting and interrelated objectives:

- To evaluate the different environmental cognitions of farmers towards selected key themes related to the concept of food security.
- To investigate and assess the environmental perceptions, attitudes and behaviours of conventional and organic farmers, in central southern England, towards organic farming and the development of more environmentally-friendly farming practices.

Global food prices, of many major food and feed commodities, have increased significantly in recent years (House of Commons, 2009). For example, during 2007 the price of many basic food staples such as wheat and rice increased by 50 and 20 per cent respectively (Chatham House, 2008:2). Very high price rises across a wide range of food commodities is unusual. Although grain prices subsequently lowered to 2006 levels, a series of violent protests and demonstrations occurred in many countries across the developing world. Estimated population increases suggest that the world population will reach nine billion by 2050 (95 per cent of this growth will occur in the developing world), thereby increasing the long-term demand for food. Peak oil prices are a key reason for recent increases in food production and distribution costs (although petroleum costs do not comprise a major proportion of energy in agricultural production (Dodson et al., 2010)), resulting in high retail food prices and, as a consequence, making it increasingly difficult to provide food security. World oil prices have reached more than $100 a barrel and, at the time of writing (May 2011), the current price is still averaging in excess of this figure (Mason, 2011). The World Bank suggests that rather than having an agribusiness-based and petrochemical-dependent industrial agriculture, a way of achieving food security is to increase productivity using GM technologies. The claimed environmental benefits of such agricultural methods relate to a reduction in existing high pesticide and fertilizer usage. However, the widespread use of GM technology might further intensify the production of monoculture crops and change some land use from food to fuel production, thereby exacerbating food security problems.

Another approach to achieving food security is to adopt the strong science-oriented, or technocentric, concept of sustainable intensification (Godfray et al., 2010). This system attempts to achieve higher yields from the same acreage without damaging the environment. Supporters of this approach claim that substantial increases in crop yield can be provided through science and technology. Examples are crop improvement, more efficient use of water and fertilizers, the introduction of new non-chemical approaches to crop protection, the reduction of post-harvest losses and more sustainable livestock (Maye & Ilbery, 2011). However, it is debatable whether sustainable intensification can be achieved without significant increases in the use of chemical inputs. Yet, such high levels of pesticide usage reduce the ecological bases of sustainable farming, thus damaging prospects of achieving food security.

A contrasting approach to conventional agriculture is organic farming, which can play a role in adapting to and mitigating the impacts of climate change. However, its role in food security debates is far from clear. Organic agriculture is a holistic production management system that promotes and enhances agro-ecosystem health, including biodiversity, biological cycles and soil biological activity. It emphasises the use of management practices in preference to the use of off-farm inputs, recognising that regional conditions require locally-adapted systems (Codex Alimentarius Commission, 1999). More recently, in March 2008, the World Board of the International Federation of Organic Agriculture Movements (IFOAM) approved the following definition:

'Organic agriculture is a production system that sustains the health of soils, ecosystems and people. It relies on ecological processes, biodiversity and cycles adapted to local conditions, rather than the use of inputs with adverse effects. Organic agriculture combines tradition, innovation and science to benefit the shared environment and promote fair relationships and good quality of life for all involved'.

An April 2008 report by the International Assessment of Agricultural Science and Technology for Development (IAASTD) recommended small-scale farmers and agro-ecological methods as the way forward in the current food crisis. Professor Bob Watson,

Director of IAASTD, claimed: 'that continuing to focus on production alone will undermine our agricultural capital ...'. A December 2010 United Nations Special Rapporteur on the Right to Food stated that: 'Moving towards sustainability is vital for future food security and an essential component of the right to food'. The report also recommended the dissemination of knowledge about the best sustainable agricultural practices. However, the concept of agricultural sustainability is a multi-faceted one involving agronomic, ecological, economic, social and ethical considerations (Farshad & Zinck, 2003) and means different things to different people (Redclift, 1987; 1992 and O'Riordan, 1997).

Water management is one of the key determinants of agricultural sustainability and therefore provision of adequate water supplies is an important requirement for the sustainability of organic and conventional farming, the UK's two principal agricultural systems. It is, however, debatable which of these two farming systems is more sustainable, although it is assumed that conventional farming will contribute most to achieving future food security. In the UK, a country which rarely experiences severe water shortages, the driest April on record (2011) resulted in the River Derwent in Cumbria being virtually dry and some reservoirs draining away. The Environment Agency stated in May that: '... if the very dry weather continues we may look at preventing farmers taking water from rivers to irrigate their crops' (Johnston, 2011). The long-term frequency and severity of such extreme climate events in the UK could have serious consequences for food security, potentially causing reduced crop yield, crop failure and farmers having to grow a different variety of crops. Agriculture is the UK's sector most affected by climate change; it also contributes greatly to climate change through the use of fertilisers, fuel and methane from ruminating livestock. Agriculture, therefore, has the greatest need for adaptation. It is also imperative to reduce greenhouse gas emissions produced by the food system, reduce dependency on fossil fuels and stop depleting natural resources such as soil and water upon which food production depends (House of Commons, 2009, p. 13). These requirements may clash with attempts to produce more food.

The five closely related food security themes discussed above (pesticides, fossil fuels, agricultural sustainability, GM crops and global climate change) are difficult to discuss in isolation as there are strong and quite complex connections between them. They can be considered a network of interrelated concepts; for example, it is almost impossible to examine the theme of food security without discussing agricultural sustainability. This crucial relationship between sustainability and food security was emphasised by Lang (2009, p. 30): 'food security can only mean sustainability'.

The rest of the chapter is divided into four sections. The next, conceptual, section outlines the key dimensions of a modified behavioural approach. This is followed by a description of the adopted two-part 'extensive' and 'intensive' research methodology used in the investigation. Section four then provides detailed insights into farmers' environmental behaviour and perceptions, attitudes and behaviours towards key food security themes, as well as appraising the consistency of farmers' environmental attitudes and behaviours. A final section provides a conclusion to the chapter.

2. A modified behavioural approach

The 'behavioural environment' - where the internal or perceptual environment in which facts of the phenomenal world are organized into conceptual patterns and given meaning or values by individuals within particular cultural contexts - was introduced into geography by William Kirk in 1952. This approach emphasizes the importance of perception in human geography, the significance of subjective experience and the potential of people as active agents in the environment. Fundamental to behavioural approaches is the idea that a crucial distinction can be drawn between the real world - the world as it is in and of itself - and the world as perceived, that is the world as humans believe it to be. The behavioural interface is the black-box within which humans form the image of their world. The schemata, or basic framework, within which past and present environmental experiences are organised and given locational meaning is the cognitive mapping process. The key psychological variables intervening between environment and human behaviour are a mixture of cognitive and affective attitudes, emotions or affective responses, perception and cognition, and learning (Golledge & Stimson, 1987).

Behavioural approaches have been used extensively in agricultural geography (Wolpert, 1964; Gasson, 1973, 1974; Gillmor, 1986; Ilbery, 1978, 1985; Brotherton, 1990; Morris & Potter, 1995; Wilson, 1996, 1997; Beedell & Rehman, 1999, 2000; Burton, 2004; Kings & Ilbery, 2010) and applied to the analysis and 'explanation' of farmers' behaviour. The focus of these approaches on individual decision makers, together with the possibility of formulating relatively 'simple' questionnaire and interview-based research methodologies, are the major reasons why behavioural approaches have been adopted by those seeking to 'understand' the decision making of farmers. Most importantly, behavioural approaches allow for the recognition of farmers as independent environmental managers who often make decisions about the management of environmental resources on their farms independent from the state or other 'official' environmental managers (Wilson, 1997).

An important aspect of the modified behavioural approach adopted in this chapter is the way in which the processes of perception and cognition influence farmers' environmental attitudes, decisions and behaviours. A specific model of environmental behaviour (a variant of the classic behavioural model) has therefore been developed to facilitate an environmental understanding of five key themes related to the concept of food security: pesticides, fossil fuels, agricultural sustainability, GM crops and global climate change (Fig 1). These closely related agri-environmental topics are associated with the working practices of organic and conventional farmers. Importantly, this type of socio-psychological framework differs from the classic behavioural approach in its focus on the concepts of perception and cognition as key parts of the decision making process.

The starting point for this conceptual framework is taken as the 'real world', which is the source of information. Knowledge is filtered through a system of perceptual receptors which are essentially the five main senses. Perception is the term given to the neurophysiological process of the reception of stimuli from an individual's surroundings (Pocock, 1974). In this process, sight is generally thought to be the major element, but other senses such as hearing and smell may also play their part. Perception is usually regarded as being immediate i.e. it follows directly upon the stimulus, and is stimulus-dependent since the nature and very presence of the perception depends on the existence and type of stimulus.

Cognition is the wider personal context of perception (Pocock, 1974). It is not necessarily immediate in the same way, since it constitutes the means of awareness that intervenes between past and present stimuli and the behavioural responses of the present and the future. The whole complex of cultural response, such as memory, experience, values, evaluation, judgement and discourse, is present in the processes of cognition. Meaning is given to information through an interaction between the individual's value system and their stored 'image' or cognitive map knowledge of the real world. The remaining filtered

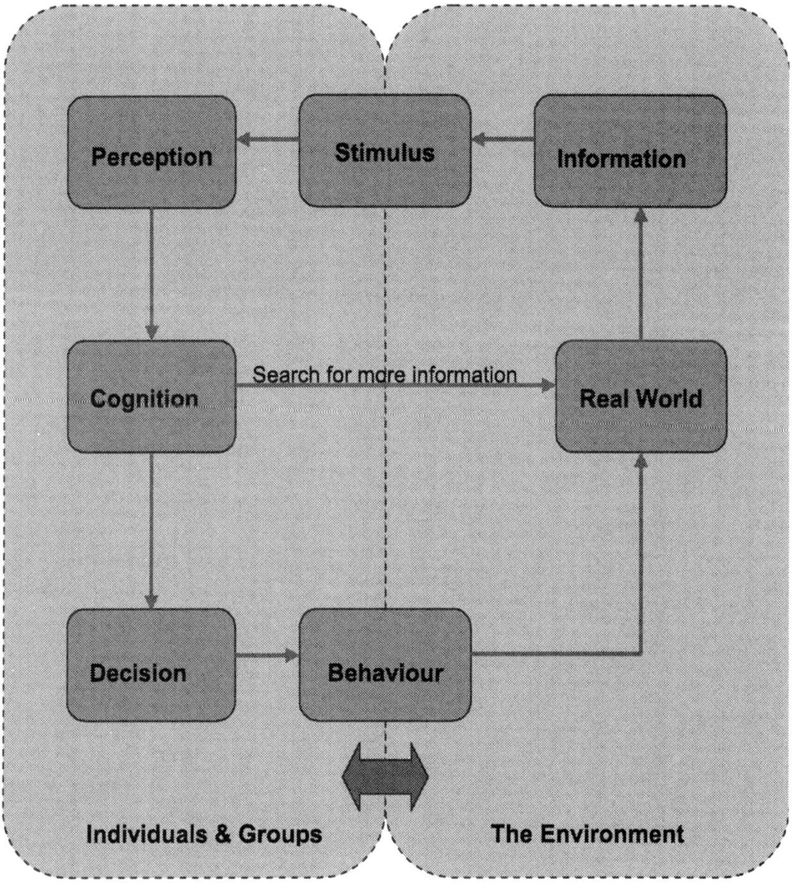

Fig. 1. A model of environmental behaviour

information is then used to update the cognitive map knowledge and to formulate a behaviour decision. This decision leads either to a reiteration of the whole process, creating another search for information from the real world until sufficient information has been acquired or some time/cost limitation acts to constrain the search, or to overt behaviour. As a result of the latter, the real world undergoes a change, fresh information becomes available and the whole process begins again.

Cognition is likely to vary from individual to individual and hence from group to group, but most such units seem to have enough in common between their cognitions to make it possible to co-ordinate thought and action (Simmons, 1993). However, there is likely to be a discrepancy between words and deeds. Cognition and perception lead to behaviour itself, which may be considered as the taking of action in regard to some environmental feature such as if, or when, to apply pesticides to a crop. There are psychological differences between individuals (leading, for example, to differences in physical sensation when exposed to the same stimuli); there are also differences in cognitive attitudes which are related to age, experience and gender. However, attitudes do not directly explain behaviour since it is possible to arrive at an attitude in a number of different ways and from different experiences. Attitudes are highly complex and therefore such a direct link is unrealistic. A simple behaviouralist explanation of decision-making is over-generalised in making the assumption that there is a basic stimulus - response in decision making (Walmsley & Lewis, 1984). In order to avoid problems associated with behavioural approaches using only inflexible structured questionnaire methodologies, and which focus on individual decision makers out of their social milieus, this study combines a balance of quantitative and qualitative work (see Burton, 2004).

3. A methodological framework

A two-stage methodology was adopted for an examination of farmers' attitudes towards the five key themes already outlined. As stage one, twenty-five organic farmers and twenty-five conventional farmers - located in central-southern England - were interviewed by telephone for about one hour. Most farmers are accessible by telephone, but may not be listed in business or private telephone directories. The first to be interviewed were organic farmers selected from the official regional Soil Association and Organic Farmers and Growers membership lists. At the end of the interview, they were asked to name a neighbouring conventional farmer, if possible of a similar size and type, who they perceived as a suitable candidate for interview. This method provided dependable geographically linked pairs of farmers for the duration of the study. Some researchers claim that a disadvantage of telephone interviewing is the problem of sample representativeness. However, within the context of this research, a 'truly' national representative sample was not anticipated, as it is limited to a specific geographical area i.e. central southern England, which may or may not be representative of farms and farmers in the UK as a whole.
A questionnaire designed in four sections was used in the 'extensive' data gathering approach. Section one contained six closed questions regarding farm size and type. Section two consisted of twenty-four open questions which explored farmers' attitudes towards farming, the environment and food security in the UK. The third section required five questions to be answered regarding farmers' specific environmental behaviour. The fourth section of six questions was aimed at eliciting personal details about the respondents. The last question asked respondents if they were willing to take part in a follow-up 'intensive' interview, and to confirm their name and address. Basic closed questions, for which response options were mutually exclusive, were included, such as gender, marital status and age, which may help in selecting individuals for future research. Closed questions regarding the type of farming system employed were included to enable examination of possible correlation between this factor and the respondents' environmental perceptions. Other questions related directly to the respondents' attitudes and behaviour in the agricultural work environment. These data were analysed both quantitatively, using summarising statistics, and qualitatively, in the form of farmers' quotations and illustrative farm cameos to emphasise the arguments being developed about organic and conventional farming. This was the most important and interesting analysis and was used to support, illustrate and broaden the impression gained from the statistics. In addition, they demonstrated similarities and differences between the two study groups being examined and gave prominence to the line of reasoning being developed.

Stage two of the methodology consisted of on-farm intensive qualitative/interpretive interviews, with five geographically linked pairs of organic and conventional farmers (selected from the sample frame used for the extensive telephone interviews) for up to 3 hours. It is important to note that the reference codes assigned to the ten respondents in section 4.3 are not (in every case) the same as those used in sections 4.1 and 4.2. The aim was not to choose a representative sample, but rather an illustrative one of different ages who farmed holdings of different sizes and systems. An interview guide was designed which prompted respondents to talk about the following range of topics related to their own farms: most productive areas, the natural environment, best wildlife areas, favourite parts and less favoured parts. The interviews were recorded using a Digital Audio MiniDisc-recorder with stereo microphone and transcribed verbatim for analysis as soon after the interviews as practicable. The data generated from this 'intensive' phase of the methodology were analysed using a textual approach relying on words and meanings, rather than statistics. Another method of analysis was to contrast and compare any interesting or unusual quotations and paraphrases made by respondents, in order to demonstrate attitudinal similarities and differences. Each interview produced contextual findings relating to the 'nature' of the respondent, thereby building up in greater depth a background picture of farmers' perceptions, attitudes and behaviours in central-southern England. The data collected were maximised to provide the broadest picture possible of farming in central-southern England. In recent years, environmental issues have become more technical and removed from everyday sensory experience, thereby posing problems with testing and analysing respondents' 'self-perceived environmental knowledge', which is further complicated by the sometimes contradictory nature of the underlying science.

In the next section, the adopted 'extensive' and 'intensive' research methodology will be used primarily to examine and gain insights into the perceptions, values, opinions and behaviours of organic and conventional farmers in relation to their awareness and understandings of agri-environmental aspects of the five key themes related to food security.

4. Investigating farmers' attitudes and behaviours

The modified behavioural approach is used first, to examine the attitudes, understandings and behaviours of organic and conventional farmers (situated in central-southern England) in relation to food security themes; second, to examine respondents' environmental behaviours; and third, to ascertain if farmers' attitudes are consistent with those expressed in sections 4.1 and 4.2.

4.1 Extensive organic and conventional farmer telephone interviews

One approach to achieving an environmentally sustainable way of producing food is organic farming (Morgan & Murdoch, 2000; Hansen et al., 2001; Lotter, 2003; Darnhofer, 2005; Kings & Ilbery, 2010). However, it seems unlikely that organic methods of food production will be adequate to provide food security in the foreseeable future. This attitude was typified by a quote from one organic respondent who farms 18 ha of arable crops: 'Absolutely, that is why I am doing it' (OF20). The average size of organic farms in the survey was 85.4 ha, which contrasted with an average size for conventional farms of 202.3

ha, although the size of both farm types was extremely variable. In contrast, conventional farmers' replies were generally not in support of those views and are typified by the following comment: 'I think organic grass farmers cause more problems with nitrates than I do, by ploughing clover [into their soil]' (CF13). Grass and fodder enterprises associated with organic livestock were the most popular organic types, occurring on a majority of the organic farms examined in the survey and any cereals grown were normally used as livestock fodder or seed. According to Willer & Gillmor (1992), it is common for farmers to experiment with organic grass production before deciding to fully convert their whole farm to organic food production. In contrast, the conventional respondents in the survey tended to grow more arable crops.

The analysis continued by listing the reasons given by organic farmers for their change to, or adoption of, an organic farming system; this is shown in descending rank order (Table 1a).

Listed in rank order by reason for adoption	Frequency	Percentage
Environmental reasons including pesticide concerns	12	48
Considered they had always farmed organically	5	20
Financial reasons including customer requirements	4	16
Not in farming previously	1	4
Small scale - needed to go intensive or organic	1	4
Had farmed organically on a previous farm	1	4
The challenge	1	4
Total	**25**	**100**

Table 1. (a) Reasons for adoption of an organic farming system

Listed in rank order by reason for non-adoption	Frequency	Percentage
Wouldn't suit the ground/way we farm	7	28
Financial reasons - producer and buyer	6	24
Cramping, restrictive and ruling out modern science	5	20
Don't think it always works	4	16
Would consider changing to organic	1	4
No, but no reason given	1	4
My farm is organic for all intents and purposes	1	4
Total	**25**	**100**

Table 1. (b) Reasons for non-adoption of organic farming

Table 1a shows that 48% of organic farmers adopted organic methods of food production because of environmental concerns such as high pesticides usage and a further 20% consider they have always farmed organically. One 45 year old owner-occupier farmer claimed: 'the toxicity of the pesticides used in my intensive agriculture made me feel quite poorly ...', demonstrating his deep concerns with health problems associated with the pesticides used in conventional farming (OF10). Of those twelve farmers who gave environmental reasons for changing to organic, three are unqualified, one has a certificate, two have a diploma, four a degree, one a higher degree and one a Doctorate; this suggests a link between higher education and environmental awareness. In contrast, Table 1b provides a list of reasons provided by conventional farmers (currently the largest contributor to food security in the UK) for their non-adoption of organic farming, shown in descending rank order. Four key findings emerge from Table 1b: first, 28% of conventional respondents said organic farming wouldn't suit their type of land; second, 24% gave financial reasons for their non-adoption; third, 20% gave technocentric reasons; and fourth, only one conventional farmer said he would consider changing to organic. It is likely that, within this study group, most conventional farmers who had the propensity to change to organic have already done so.

Analysis proceeded by asking respondents what they know about the amount of fossil fuels used by some organic farmers in the mechanical weeding processes. The key finding is that almost all the organic respondents said organic arable farmers use a lot of fossil fuels. This practice contributes to climate warming and is therefore liable to have a detrimental effect on food security through reduced crop yield and/or failure. However, several organic respondents declined to comment possibly because they were aware that more fossil fuels are used in mechanical weeding processes than chemical methods of weeding. Most conventional farmers in the survey were not critical of organic farmers in relation to this issue.

The diversity of farmers' attitudes and cognition in relation to whether organic agriculture is an environmentally sustainable method of food production is demonstrated in Figure 2.

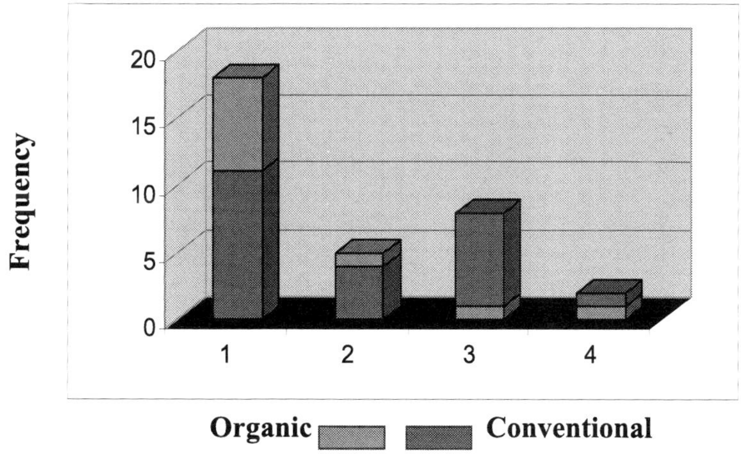

1. Sustainable. 2. Could be sustainable. 3. Not sustainable. 4. Don't know.

Fig. 2. Frequency of farmers' opinions on sustainability of organic farming

Unsurprisingly, more organic (72%) than conventional farmers (44%) felt that organic agriculture is a sustainable form of food production. Another 20 per cent of organic respondents and 16 per cent of conventional farmers thought that it could be. Agricultural sustainability is crucial for maintaining long-term food security. Thirty-two per cent of conventional farmers said that organic food production is unsustainable in comparison with only four per cent of organic farmers.

Both survey groups were asked if they thought conventional food production was having negative environmental impacts. The frequency of farmers' responses have been categorised under the five headings graphically detailed in Figure 3.

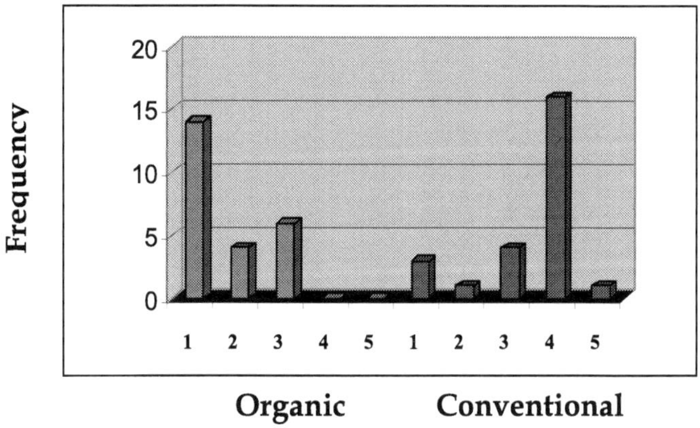

1. Yes
2. Yes with specific reason
3. Not always/depends on farmer
4. No
5. No with specific reasons

Fig. 3. Frequency of farmers' response to the environmental impacts of conventional food production

The figure shows that fewer conventional than organic respondents perceive conventional food production to have negative environmental impacts. Fourteen (56%) organic farmers thought that conventional agriculture is harmful to the environment in contrast to three (12%) conventional farmers. Another four (16%) organic farmers said yes, but qualified their answer with specific reasons for their environmental concern such as: 'high levels of nitrate and pesticides on crops'. High pesticide usage can lead to further pest resistance and farmers' reliance on agribusiness, thereby creating significant environmental costs and adversely affecting food security. Only one organic farmer (OF3) thought conventional food production is not having negative impacts compared with 16 (64%) of conventional farmers who agreed with that statement. One 45 year old owner-occupier organic farmer with a post graduate degree replied: 'not always, like anything at its worst it depends on the farmer and his attitude …' (OF10). Earlier analysis showed that, although the two survey groups were relatively evenly matched in the number of vocational qualifications, organic farmers have more qualifications towards the upper end of the education spectrum and it is only organic farmers who have a higher degree. Dunlap et al., (2000) found that environmental concern was correlated with variables such as age and education. However, a different picture is revealed if the qualifications relating specifically to agriculture are examined i.e. more conventional than organic farmers have a national certificate in agriculture.

Analysis shows that the organic farmers in the survey are very critical of pesticide use. This contrasts with conventional farmers, who said they need to use pesticides to produce their crops but, nevertheless, are aware of the dangers of over-use of such chemicals. One 54 year old owner/tenant organic farmer said: 'If you read the magazine that the conventional farmers read, the *Farmers Weekly*, you will notice that the magazine is paid for by pesticide adverts. The biggest adverts saying this is the time to spray with this or that. This is the way [conventional] farmers are being educated ...' (OF5). Some conventional farmers agreed with organic farmers that conventionally produced crops sometimes use high levels of pesticides, which is an unsustainable method of seeking continual productivity gains to mitigate food insecurity. Three recent independent studies, carried out in the USA, found that children whose mothers are exposed to common agricultural pesticides are more likely to experience a range of deleterious effects in their cognitive development, including lower IQ, as well as impaired reasoning and memory (Eskenazi, 2011).

Many of the organic and conventional respondents perceive GM food production as an integral part of conventional agriculture. This view is supported by Lawrence et al. (2010). GM crops were investigated in two broad, but overlapping, categories: first; those concerning environmental issues; and second, those which relate to public health concerns. More conventional farmers have technocentric attitudes and a greater acceptance of GM crops than the organic respondents in the survey, typified by organic farmers' comments: 'I think it's tampering with nature ...' (OF19). Most conventional farmers in the survey seemed more accepting of GM technology, typically saying: 'I haven't a huge fear of them as long as we observe the science ...' (CF24). However, it may be unwise to believe that science and technology are a panacea, as new technologies often raise further questions and complications of their own (Frow et al., 2009). Conventional farmers are usually less critical of GM crops than the organic respondents, and seem to place their main emphasis on the potential environmental benefits to be gained from a reduction in pesticide usage. But, as discussed earlier, greater use of GM technology is likely to further intensify the production of monoculture crops and change some land use from food to fuel production, thereby jeopardising food security.

Analysis reveals that more organic than conventional farmers have concerns about the potential health risks associated with eating GM food. Other organic respondents have some misgivings and perceive that there will be future public health concerns. However, a number of organic farmers could not think of any health issues. Generally, the conventional farmers seem to have fewer health concerns relating to GM food. A key finding is that more conventional than organic farmers believe that GM farming will be necessary to feed a growing world population. This could help alleviate problems of food insecurity, but with the loss of some agricultural biodiversity and therefore sustainability. A number of organic farmers have concerns about GM crops epitomised by the following: 'It's [GM] not necessary [to feed a growing world population] - it's an argument used by the chemical companies' (OF25).

Global climate change (with increasing frequency of extreme weather events) is a particularly important issue to many UK farmers who perceive that their future livelihood will be endangered by crop reduction/failure causing food insecurity. In July 2011, the Energy Secretary Chris Huhne agreed: 'A changing climate will imperil food, water, and energy security...' (Anon, 2011). Climatic change is characterised by a rapid increase in global temperature and is changing at an unprecedented rate (Pulido & Berthold, 2006). The evidence for such rapid recent climate change is now compelling (IPCC, 2007). A number of conventional farmers perceived that changes in weather patterns are part of the normal course of events. In contrast, some organic respondents said that global warming was primarily caused by burning fossil fuels but, as shown earlier, organic farmers admit to using large amounts of fossil fuels, thereby contributing to climate change. Interestingly, only one respondent (OF1) specifically referred to agriculture affecting climate change although, as discussed in Section 1, agriculture is the UK's sector most affecting climate change. In common, all respondents said that they may have to grow a different variety of crops if the climate gets warmer and some referred to the expansion of growing maize in the UK as evidence of global warming taking place.

This section has demonstrated significant differences, but also similarities in the perception and cognition of members of the two survey populations in relation to environmental aspects of the five closely related food security themes. It proved difficult to discuss any of the five food security themes in isolation as there are strong and quite complex connections among many of these issues. Examination of the environmental behaviour of the same two groups of farmers carries forward the analysis in the following section.

4.2 Farmers' environmental behaviour

To gain a greater understanding of farmers' perceptions, attitudes and values, this section examines the respondents' environmental behaviour in the countryside. Again, these relatively simple data are supported by qualitative data in the form of farmers' quotations. An important part of this analysis is concerned with the way in which organic and conventional farmers make sense of environmental issues through the processes of perception and cognition.

Analysis proceeds in two stages: first, respondents' membership of agri-environmental schemes, including their participation in conservation work and membership of environmental organisations; and second, their 'readership' of agri-environmental journals and magazines and how they believe farmers should behave in the countryside. A number of key differences emerged between the two survey groups.

Seventeen (68%) organic and thirteen (52%) conventional farmers participate in agri-environmental schemes; the remainder of respondents do not belong to any schemes. Two key points emerged from the analysis: first, considerably more organic than conventional farmers are members of an agri-environmental scheme; and second, three times more organic than conventional farmers belong to more than one scheme. Organic farmers' high membership of agri-environmental schemes supports the finding shown in Figure 2 - that organic respondents perceive organic agriculture to be an environmentally sustainable means of food production, although such methods are unlikely to mitigate problems of food insecurity. Over half of the organic farmers were in the Countryside Stewardship Scheme, contrasting significantly with just over one tenth of conventional farmers. Until the launch of Environmental Stewardship, Countryside Stewardship was the government's principal scheme. Farmers entered 10-year agreements to manage their land in an environmentally friendly way in return for annual payments (DEFRA, 2002). There were equal numbers of both types of farmer involved with the set-aside scheme. The European Union (EU) introduced set-aside of arable land in 1988 as part of a package of measures designed to reduce over-production in member states to allow reductions in the costs of agricultural price support (Floyd, 1992). In contrast to organic farmers' high membership of Countryside Stewardship, there were twice as many conventional as organic farmers involved with the ESA scheme. Similar to Countryside Stewardship, government offered financial incentives to farmers and other land managers who agree to undertake environmentally beneficial practices under the ESA scheme.

Further significant differences between organic and conventional farmers were found in relation to carrying out conservation work. First, more organic than conventional farmers undertake conservation work, with a much higher proportion involved with hedge laying and wood planting. Secondly, conventional farmers see the creation of pheasant cover as conservation works. Carr & Tait (1991, p. 286) found that conventional farmers tended to perceive pheasants as a: 'wildlife species beneficial to farming'. The organic farmers' conservation behaviours support their environmental attitudes shown in Table 1a which provided environmental reasons for their adoption of 'sustainable' organic methods of food production although, as discussed previously, such methods are unlikely to provide a permanent solution to food insecurity.

Both groups of respondents were asked if they were members of any environmental organisations such as their local wildlife trust. Their responses are listed in rank order in descending frequency of mention by both groups of farmers (Table 2). The table shows that less than half of both categories of farmers are members of environmental institutions and organic farmers prefer the Wildlife Trust, Woodland Trust and Friends of the Earth, whereas conventional farmers prefer the Game Conservancy Trust. The same number of organic as conventional farmers claims to be members of environmental organisations. However, this raises the important question about what an environmental organisation is and just how 'green' their credentials are. For example, Table 2 shows that the Game Conservancy Trust was the most frequently mentioned agency by respondents. However, of those, over three quarters were conventional farmers which, when cross-tabulated with their main countryside leisure pursuit, were found to be shooting. The Game Conservancy Trust, now renamed - the politically more acceptable - Game & Wildlife Conservation Trust (GWCT) claims to be the leading UK charity conducting scientific research to enhance the British countryside for public benefit. A recent impartial report of Grouse shooting commented: 'The bloodlust is extraordinary. I sense it in myself, but it is obvious in these alpha males with their fingers on the trigger' (Hollingshead, 2010). Organic farmers' membership of 'truly green' environmental organisations is consistent with their concerns regarding the sustainability and health issues related to the food security themes, pesticides and GM crops.

When examining these results in relation to the readership of magazines and journals, it was found that a total of 27 different periodicals were mentioned. *Farmers Weekly* and *Farmers Guardian* dominate conventional farmers' reading. The most popular magazine with interviewees is *Farmers Weekly*, which is 'read' by just over three quarters of organic farmers and almost all conventional farmers. Earlier in the analysis, OF5 criticised *Farmers Weekly* for influencing the quantity of pesticides used by conventional farmers. The second most popular magazine 'read' by over a quarter of conventional farmers - and no organic farmers - is *Farmers Guardian*. The *Living Earth* and *Organic Farming* were read by 88 per cent of organic farmers but by no conventional respondents. Similarly, no conventional farmers read *The Ecologist* which was read by 8 per cent of organic respondents. The last three periodicals have a strong focus on rejection of pesticide use and GM technology which have adverse effects on agricultural sustainability and food security. Organic farmers read much more widely and seemed more critical in their reading tastes than conventional farmers which may be related to organic farmers on average having higher academic qualifications than the conventional respondents.

	Organic farmers		Conventional farmers	
Descending rank order in Frequency of membership	**Frequency**	**Percentage**	**Frequency**	**Percentage**
Game Conservancy Trust	2	8	8	32
Wildlife Trust	5	20	1	4
Woodland Trust	4	16	0	0
National Trust	3	12	3	12
Friends of the Earth	3	12	0	0
RSPB	2	8	2	8
Wildfowl and Wetland Trust	1	4	0	0
Greenpeace	1	4	0	0
Rare Breeds Survival Trust	1	4	0	0
FWAG	0	0	1	4
LEAF	0	0	1	4
None	14	56	14	56
Total participating	**11**	**44**	**11**	**44**

Table 2. Frequency of farmers' membership of environmental organisations

Respondents were asked the loosely worded question about how they thought farmers should 'behave' in the countryside, to ascertain which agri-environmental issues they perceive as important. The answers were analysed by means of using the 'frequency of mention' of key environmental/agricultural words/concepts used as indicators of the respondents' level of environmental behavioural awareness. These words/concepts are listed in an order loosely linked to adjacent themes in order to gain a progression of ideas throughout the analysis (see Table 3). Four key findings emerge from the analysis: first, the term 'behave responsibly' is used by over a quarter of organic farmers and somewhat fewer conventional farmers; second, almost a quarter of organic farmers, contrasting with less than a tenth of conventional farmers, use the words 'stewards, keepers, custodians or protectors'; third, the same number, in both survey groups, use the terms 'pride, respect or sensitivity towards the environment'; and fourth, more conventional than organic farmers use the words looking after, care and good condition. Interestingly, only a fifth of conventional farmers referred to 'producing food' contrasting with fewer organic farmers' specific reference to the 'quality' and 'locality' of food produced. These comments suggest that although organic food may be available, it could be too expensive for increasing numbers of low-income families who will possibly experience some food insecurity.

	Organic farmers		Conventional farmers	
Key agri-environmental words	**Frequency**	**Percentage**	**Frequency**	**Percentage**
Behave responsibly	7	28	4	16
Stewards/keepers/custodians/protectors	6	24	2	8
Pride/respect/sensitivity to environment	6	24	6	24
Looking after/care/good condition	4	16	5	20
Future generations/family/children	2	8	4	16
Community	0	0	1	4
Quality of life	0	0	1	4
The farm in its own right	2	8	2	8
Sustainability	1	4	1	4
Environmentally friendly	1	4	1	4
Quality of food/local food	2	8	0	0
Behave the way organic farmers behave	1	4	0	0
As they have always behaved	0	0	2	8
Produce food/customer/right job/a living	0	0	5	20
Open minded	2	8	0	0
Do more for their image	0	0	1	4
Shoot on site	1	4	0	0
Inappropriate answers	1	4	1	4
Can't think of an answer	2	8	2	8
Total	**23**	**92**	**23**	**92**

Table 3. Frequency of mention of key-environmental words

This section has demonstrated some significant differences, similarities and overlap in environmental behaviours between members of the two survey populations. To add depth to the analysis, the final stage of the research explores environmental aspects of the five related environmental farm themes by means of intensive on-farm interviews with selected conventional and organic farmers.

4.3 On-farm qualitative interviews

This section seeks to ascertain if respondents' understandings of the five food security themes, and their environmental behaviour, are compatible with their cognition of their own farm environments. This is achieved through discussion of a range of related farm topics listed below:

- Most productive areas
- The natural environment
- Best wildlife areas
- Favourite parts
- Less favoured parts

A number of similarities and differences emerged between case study respondents' understandings of these associated themes, when related to their cognition of the five core themes examined in section 4.1 and their environmental behaviour detailed in section 4.2. All conventional farmers said that some parts of their farm are more productive than others and related productivity to practical issues such as field size, their farming practices, relative field heights, drainage, 'natural' differences in soil fertility and, in some instances, the quantity and type of fertilisers used. In comparison, several organic farmers seemed proud of their soil's lack of productivity as it produces an abundance of wild flowers. Although soil fertility was not measured, respondents perceived considerable variability within and between fields on farms of different sizes, types and altitudes. Although in this limited survey sample the conventional respondents are on average less well qualified academically than the organic farmers, CF1 and CF4 equated soil fertility to technical issues such as field heights combined with high levels of fertilisers and the resulting crop income. This contrasted with discussions of the 'natural' differences in soil fertility (OF2) and pride in lack of fertiliser use (OF1). Additionally, the discussions revealed significant diversity of opinion to the resulting plant species due to differences in field productivity. For example, wild flowers were thought of as weeds by (CF2) contrasting with the chalk-loving wild flowers of (OF1). Conventional farmers tended to optimise yields (to maximize their income) which is important for maintaining food security, whereas, the organic respondents seemed to place less importance on this issue. This is likely to be related to some of the organic farmers having off-farm income.

Discussing natural environment aspects of respondents' farms revealed significant differences in focus and cognition between the 'two' farmer types. For example, most conventional respondents associated the natural environment with creating suitable conditions for hunting and/or shooting. But, in accord with most organic respondents, (CF1) professes to see the natural and farmed environments as one and the same and believe that it is important for someone to 'own' the land. Table 3 showed that twice as many conventional as organic farmers are concerned with future generations and children. This contrasts with three times as many organic as conventional farmers' concern with being 'stewards' or 'custodians' of the countryside. Most organic farmers saw a direct relationship between the natural and farmed environment and emphasised the conservation work they have carried out to increase their farm's biodiversity, thereby positively influencing the sustainability of their agricultural food production system. Some conventional respondents place equal importance upon a range of what would seem to some organic respondents as irreconcilable and conflicting countryside issues, such as looking after the landscape, giving access to the public, looking after pheasants, hunting and shooting. However, some diversity of focus and understanding was shown within the conventional farmer group.

Hedgerows were considered one the best areas for wildlife by most conventional farmers. All references made about hedges by conventional respondents were regarding cost, maintenance, or lack of, in contrast to some organic farmers' reference to planting these linear strips of woodland which are important for increasing the biodiversity and sustainability of the agri-environmental food production system. The value of the whole farm for wildlife was also made by respondents OF5, OF3 and OF2 who supported such viewpoints with examples of specific farm habitats with their associated mammals, birds and invertebrates. Respondents' comments demonstrate a continuum of environmental attitudes, ranging from mixed (CF3), who claims there is no difference in environmental quality in various parts of his land, to mixed (OF1), who emphasises the total biodiversity of his farm holding. These attitudes may be related to CF3 leaving school without formal qualifications and gaining his 'education' 'on the job'. This contrasts with OF1 who, although his 'A' levels were too weak to enter university, gained a Diploma in Agriculture and has since pursued an academic interest in specific research on his farm.

While discussing favourite parts of the farm, one farmer focused his observations, and indeed anger, on one of his neighbour's - the Woodland Trust - lack of action and illustrated his extreme tidy-ness by passing adverse comments on their policy of allowing fallen trees to remain on the woodland floor (CF1). The Woodland Trust is a conservation charity concerned with the protection and sympathetic management of native woodland heritage. Such fallen dead timber in the food chain becomes resources for other organisms such as decomposers (bacteria and fungi) thereby ultimately aiding biodiversity, agri-environmental sustainability and food security. The respondent's tidiness is in line with the findings shown in Table 3, that more conventional than organic farmers use the words looking after, care and good condition of their land. He uses words like 'production' to describe wildlife which is a word that suggests the process of being manufactured, especially in large quantities, rather than natural processes, thereby further revealing his technocentric attitudes and criticises his neighbours for their lack of management [shooting] of woodland 'vermin'. Other conventional respondents emphasised their appreciation of the isolation and tranquillity of some less accessible parts of their farms. In contrast, two organic interviewees were very specific about their favourite areas of their farms and focused on environmental aspects of wildlife sites and the beauty of those habitats thereby demonstrating their agri-environmental awareness. This discussion has revealed a range of differences between interviewees' views of their favourite farm parts, which provided further insights into their agri-environmental attitudes and behaviours by demonstrating what is important to them.

Discussing least favourite farm parts revealed some similarities and differences in the attitudes of the respondents. Some conventional interviewees associated the term 'least favourite' with specific farm problems such as changing his stocking regime from cows to sheep. Equally pragmatic, OF5 dealt with her problem of a poor quality field by disposing of it to another local landowner. In contrast, OF1 said that he found almost all parts of his land pleasing; however, he did comment that his arable fields were the most boring parts of the farm holding, possibly due to his enthusiasm and focus on his wildlife habitats. This discussion has shown similarities in the focus of some respondents, such as the practical approaches of CF1, CF4 and OF5 in dealing with less favoured parts of their farms. In contrast, OF1 claims to favour all parts of his land thereby suggesting an appreciation of his farm's biodiversity which, as discussed earlier, is crucial for the sustainability of food production and food security.

A diverse range of attitudes and behaviours emerged from the farmers' discussions - whether loosely labelled conventional or organic - when asked to talk about the five related themes, thereby providing some insight into their agri-environmental perceptions, attitudes and behaviours relating to biodiversity, sustainability and food security. For example, the comments of mixed tenant farmer (CF1) revealed his business-like attitude, tidiness and technocentric nature towards food production and his propensity for hunting. However, some of his farm behaviours are at variance with his opinions, as exemplified by his criticism of large fields when he manages the largest, and still growing, farm holding in the survey. However, it is common for there to be no simple relationship between verbal and non-verbal indicators of an attitude. Such extremely large fields reduce crop diversification by relying on planting of monocultures over large areas and reduce biodiversity by excluding many species which may otherwise have been present. In contrast, mixed tenant farmer (OF1) has a less tidy approach towards farming and seems to have a more ecocentric attitude and a very different cognition of, and relationship with, the farm environment and food production, exemplified by the use of terms such as 'loving hay meadows'.

5. Conclusion

This chapter aimed to compare the cognition, attitudes and behaviours of farmers loosely labelled 'organic' and 'conventional', in central southern England, especially in relation to farming, the environment and five key themes related to the concept of food security. Using a modified behavioural conceptual framework revealed a spectrum of agri-environmental perceptions, attitudes and behaviours among farmers in relation to five key food security themes. However, farmers' behaviours cannot be directly explained by their attitudes because it is possible to arrive at an attitude in a number of different ways and from different experiences. Attitudes are highly complex and therefore such a direct link is unrealistic.

The differences detected between the survey populations were epitomised by some conventional farmers' high levels of pesticide usage, concern with keeping the land in suitable condition for growing crops, their belief in the necessity of conventional farming methods to feed a growing world population, an anthropocentric acceptance of GM crops and the belief that changes in weather patterns are part of the normal course of events. This contrasted significantly with many organic farmers' more ecocentric approach to agri-environmental issues and belief in the need for a biodiverse and sustainable countryside whilst, at the same time, producing locally grown and consumed healthy foods which should be able to accommodate population increases. The organic respondents also had concerns about the potential health risks associated with GM crops and the belief that global climate change is principally caused by burning fossil fuels. Unsurprisingly, more organic than conventional farmers said that organic agriculture is a sustainable form of food production which, if correct, will help mitigate food security problems. Other researchers also describe organic farming as a more sustainable method of agricultural food production than most conventional farming systems (Lampkin et al., 1999; Grey, 2000; Edwards-Jones & Howells, 2001; Michelsen, 2001; Mader et al., 2002). In contrast, some writers have raised concerns that organic farming is itself becoming conventionalised (Buck et al., 1997; Tovey, 1997). But, there are different types of organic farming systems; for example, commercial organic food production has less environmental benefit than organic farming methods practised on a small scale by philosophically committed farmers. Although organic production methods are considered a useful way of reducing the current impact of agri-food production systems (Lockie et al., 2006; Schahczenski & Hill, 2009; Scherr & Sthapit, 2009), it seems unlikely that peak oil prices, combined with the need to reduce dependency on fossil fuels, will be helped greatly by organic methods of food production. Some researchers claim that organic production methods have out-performed productivist approaches by providing environmental benefits such as water retention and improved soil fertility, thereby reducing the impact of agri-food production systems on the environment (Altieri, 1998; Environmental News Service, 2009). In contrast, it has been argued that abandoning productivist methods of food production will increase global food insecurity, resulting in millions of people dying of starvation (Avery, 1995).

In common, all respondents said that they may have to grow a different variety of crops if the climate gets warmer and some referred to the expansion of growing maize in the UK as evidence of global warming taking place. However, maize is not the greenest biofuel in terms of CO_2 emissions reduction (International Energy Agency, 2004) and, used as biofuel, puts food security at further risk of leaving less food for human consumption. But climate has always changed and it is likely to do so in the future. Some researchers (more in line with the cognition of conventional farmers) suggest that fear of global warming derives from politics and dogma rather than scientific proof (Plimer, 2009).

Examination of farmers' environmental behaviour also revealed some interesting differences between the two survey populations. For example, conventional farmers are less interested in joining environmental schemes than organic farmers but, significantly, more organic that conventional farmers belonged to more than one scheme. More organic than conventional farmers carry out conservation work such as hedge laying and wood planting contrasting with conventional farmers who see creation of pheasant cover as conservation works. Membership of environmental institutions was not high among either group, with conventional farmers preferring the Game Conservancy Trust while organic farmers preferred the Wildlife Trust, Woodland Trust and Friends of the Earth. Further significant differences between organic and conventional farmers were found in relation to the readership of magazines and journals. Thus while *Farmers Weekly* and, to a much less extent, *Farmers Guardian* dominate conventional farmers' reading, the *Living Earth* and *Organic Farming* were the most widely read among organic farmers. The most popular magazine overall was *Farmers Weekly*, but organic farmers read more widely and seemed more critical in their reading habits than conventional respondents. The most significant difference between the two groups of respondents is that almost two thirds of conventional farmers shoot regularly (if only what they perceive as vermin), contrasting with less than one third of organic farmers. In response to the loosely worded question about how they thought farmers should 'behave' in the countryside, the term 'behave responsibly' was used more by organic than conventional farmers. Organic farmers also tended to use the words 'stewards, keepers, custodians or protectors', in contrast to conventional farmers who preferred to use the words 'looking after, care and good condition'. As a rule, organic respondents' agri-environmental behaviour, such as high membership of environmental organisations and participation in conservation work, supported their ecocentric attitudes expressed about the five key food security themes.

For the most part, the on-farm qualitative interviews supported the findings from the two previous sections. For example, conventional farmers tended to optimise yields using chemical inputs in order to maximize their income which, although important for achieving food security, also has damaging effects on agricultural sustainability and is therefore simultaneously detrimental to food security. This contrasted significantly with some organic farmers' pride in lack of fertiliser use and what they perceived as natural difference in their soil.

Advocates of organic farming systems - which receive substantial financial support in the form of subsidy payments - claim they are 'sustainable' and see them as a potential solution to the continued loss of biodiversity. Contrary to many published studies, however, it remains unclear whether such 'holistic' whole-farm approaches, exemplified by organic farming systems, provide such benefits for biodiversity due to the lack of longitudinal studies to 'fully' appraise their potential role as sustainable producers of healthy nutritious food. However, throughout the three stages of the analysis, generally the perceptions, attitudes and behaviours of the organic farmers in the survey demonstrated an ecocentric approach to the environment, farming and food production. Nevertheless, some organic respondents admitted using high levels of fossil fuels thereby contributing to global climate change, which subsequently has an adverse effect on food security. If in the future, however, an alternative renewable form of energy - not dependent on oil - could be found for agricultural use, then organic food production would seem appropriate to provide sustainable farming in the UK. Contrasting significantly, conventional farmers' more anthropocentric attitudes and behaviours towards producing food using high levels of pesticides cause a reduction in agricultural biodiversity thereby putting at risk the long-term sustainability of food security. One method of assuaging food insecurity is to increase the area of land under cultivation to include 'unproductive' or 'marginal land' such as set-aside (prior to abolishment in 2008); however, such thinking overlooks the important contribution set-aside makes to agricultural sustainability. Land available for cultivation is a key limiting factor for achieving food security as arable land per person shrank 40 per cent from 0.43 ha in 1962 to 0.26 ha in 1998 (FAO, 2003). In contrast, some researchers claim that organic agriculture has the potential to produce enough food on a global per capita basis to sustain the human population without increasing the agricultural land base (Badgley et al., 2007). The challenge of ensuring food security for a growing population is to produce sufficient food in a more sustainable way using resources less exploitatively, while simultaneously minimising detrimental environmental impacts such as greenhouse gas emissions.

The modified behavioural approach used in this chapter has helped to provide an awareness, sensitivity and understanding of farmers' behaviour in their geographical world. However, this was not achieved without problems such as the discrepancies experienced between respondents' attitudes and their actual farm behaviour. The research provides a conceptual and empirical contribution to geographical study and knowledge regarding the environmental perceptions, attitudes and behaviours of farmers in central-southern England.

6. Acknowledgements

We wish to record our thanks to the twenty five organic and twenty five conventional farmers who took part in the 'extensive' telephone interviews. We are especially grateful for the dedication, interest and insights provided by the five geographically linked pairs of organic and conventional farmers who participated in the on-farm 'qualitative' interviews.

7. References

Anonymous (2011) Climate change a threat to our national security, says Huhne. *The Daily Telegraph* 8th July 2011.

Altieri, M. (1998) *Ecological impacts of industrial agriculture and the possibilities for truly sustainable farming.* In: Magdoff, F., Buttel, F. & Foster, J. Hungry for Profit: Agriculture, Food and Ecology. Monthly Review Press, New York.

Badgley, C., Moghtader, J., Quintero, E., Zakem, E., Jahi Chappell, M., Aviles-Vazquez, K., Samulon, A & Perfecto, I. (2007) Organic agriculture and the global food supply. *Renewable Agriculture and Food Systems,* Vol. 22, pp. 86-108.

Avery, D. (1995) *Saving the Planet with Pesticides and Plastics: The Environmental Triumph of High-Yield Farming.* Hudson Institute, Indianapolis.

Beedell, J. D. C. & Rehman, T. (1999) Explaining farmers' conservation behaviour: Why do farmers behave the way they do? *Journal of Environmental Management*, Vol. 57, pp. 165-176.

Beedell, J. D. C. & Rehman, T. (2000) Using social-psychology models to understand farmers' conservation behaviour. *Journal of Rural Studies*, Vol. 16, pp. 117-127.

Brotherton, I. (1990) Initial participation in UK Set-Aside and ESA schemes. *Planning Outlook*, Vol. 33, pp. 46-61.

Brunori, G. & Guarino, A. (2010) *Security for Whom? Changing Discourses on Food in Europe in Times of a Global Food Crisis*. In: Lawrence, G., Lyons, K. & Wallington, T. Food Security, Nutrition and Sustainability. Earthscan, London.

Buck, D., Getz, C. & Guthman, J. (1997) From farm to table: the organic vegetable commodity chain of northern California. *Sociologia Ruralis*, Vol. 37, pp. 3-19.

Burton, R. J. F. (2004) Reconceptualising the 'Behavioural' approach in agricultural studies: a sociopsychological perspective. *Journal of Rural Studies*, Vol. 20, pp. 359-371.

Carr, S. & Tait, J. (1991) Differences in Attitudes of Farmers and Conservationists and their Implications. *Journal of Environmental Management*, Vol. 32, pp. 281-294.

Chatham House (2008) *Rising food prices: drivers and implications for development*. A Chatham House Report, Chatham house, London.

Codex Alimentarius Commission (1999) *What is organic agriculture?* (FAO/WHO Codex Alimentarius Commission, 1999).

Darnhofer, I. (2005) Organic Farming and Rural Development: Some Evidence from Austria. *Sociologia Ruralis*, Vol. 45, pp. 308-323.

DEFRA (2002) Countryside Stewardship Scheme (CSS).

Dodson, J., Sipe, N., Rickson, R. & Sloan, S. (2010) *Energy Security, Agriculture and Food*. In: Lawrence, G., Lyons, K. & Wallington, T. Food Security, Nutrition and Sustainability. Earthscan, London.

Dunlap, R. E., Van Liere, K. D., Mertig. A. G. & Jones, R. E. (2000) Measuring endorsement of the New Ecological Paradigm: a revised NEP scale. *Journal of Social Issues*, Vol. 56, pp. 425-442.

Edwards-Jones, G. & Howells, O. (2001) The origin and hazard of inputs to crop protection in organic farming systems: are they sustainable? *Agricultural Systems*, Vol. 67, pp. 31-47.

Environmental News Service (2009) *The environmental food crisis: A crisis of waste*, Environmental News Service, www.ens-wire.com/ens/feb2009/2009-02-17-01.asp.

Eskenazi, B. (2011) *Studies identify link between prenatal pesticides exposure and development in children*. Organic Trade Association, 2011.

Farshad, A. & Zinck, J. A. Seeking agricultural sustainability. *Agriculture, Ecosystems & Environment*, Vol. 47, pp. 1-12.

Floyd, W. D. (1992) Political *aspects of set-aside as a policy instrument in the European Community*. In: J. Clarke (ed) Set-Aside British Crop protection Council Monograph, Vol. 50, pp. 13-20.

FOA (2003) *Agriculture, food and water*. Rome, food and agriculture organization chapter 2, http://www.fao.org/DOCREP/006/Y4683E/y463e06.htm.

Frow, E., Ingram. D., Powell, W., Steer, D., Vogel, J. & Yearley, S. (2009) The Politics of plants. *Food Security*, Vol. 1, pp. 17-23.

Gasson, R. (1973) Goals and values of farmers. *Journal of Agricultural Economics,* Vol. 24, pp. 521-537.

Gillmor, D. (1986) Behavioural studies in agriculture: goals, values and enterprise choice. *Irish Journal of Agricultural Economics and Rural Sociology,* Vol. 11, pp. 19-33.

Godfray, C. J., Crute, I., Haddad, L., Lawrence, D., Muir, J. F., Nisbett, N., Pretty, J., Robinson, S., Toulmin, C. & Whiteley, R. (2010) The future of the global food system, *Phil. Trans. R. Soc. B,* Vol. 365, pp. 2769-2777.

Golledge, R. G. & Stimson, R. J. (1987) *Analytical Behavioural Geography.* Croom Helm, London.

Grey, M. (2000) The industrial food stream and its alternatives in the United States: An introduction. *Human Organization,* Vol. 59, pp. 143-150.

GWCT (2007) *Our History.* Game and Wildlife Conservation Trust, Fordingbridge.

Hansen, B., Alroe, H. F. & Kristensen, E. (2001) Approaches to assess the environmental impact of organic farming with particular regard to Denmark. *Agriculture, Ecosystems and Environment,* Vol. 55, pp. 11-26.

Hollingshead, I. (2010) True glory of a Twelfth on the moor. *The Daily Telegraph* 14th August 2010.

House of Commons Environment, Food and Rural Affairs Committee (2009) *Securing food supplies up to 2050: the challenges faced by the UK,* Fourth Report of Session 2009-09, Vol. 1, House of Commons, London.

Ilbery, B. W. (1978) Agricultural decision-making: a behavioural perspective. *Progress in Human Geography,* Vol. 2, pp. 448-466.

Ilbery, B. W. (1985) Factors affecting the structure of horticulture in the Vale of Evesham, UK: a behavioural interpretation. *Journal of Rural Studies,* Vol. 1, pp. 121-133.

International Assessment of Agricultural Science and Technology for Development (2008) *Urgent changes needed in global farming practices to avoid environmental destruction.* Press material from IAASTD.

International Energy Agency (2004) *Biofuels for transport: An international perspective.* Report. http://www.iea.org/textbase/nppdf/free/2004/biofuels2004.pdf.

International Federation of Organic Agriculture Movements (2008) Press release 22nd January 2010.

IPCC (2007) *Climate Change 2007: the Physical Science Basis.* Contribution of Working Group 1 to the Fourth Assessment Report of the Intergovernmental Panel on Climate Change. Cambridge University Press, Cambridge.

Johnston, P. (2011) If the Romans could do it, why can't we? *The Daily Telegraph* 5th May 2011.

Kings, D. & Ilbery, B. (2010) The environmental belief systems of organic and conventional farmers: Evidence from central-southern England. *Journal of Rural Studies,* Vol. 26, pp. 437-448.

Kirk, W. (1952) Historical geography and the concept of the behavioural environment. *Indian Geographical Journal,* Silver Jubilee, Vol. 1, pp. 52-160.

Laing, T. (2009) *How new is the world food crisis? Thoughts on the long dynamic of food democracy, food control and food policy in the 21st century,* paper presented to the Visible Warnings: The World Food Crisis in Perspective conference, April 3-4, Çornhill University, Ithaca, NY.

Lampkin, N. et al., (eds) (1999) The policy and regulatory environment for organic farming in Europe. *Organic farming in Europe: Economics and policy,* Vol. 1, University of Hohenheim, Stuttgart.

Lawrence, G., Lyons, K. & Wallington, T. (2010) *Food Security, Nutrition and Sustainability.* Earthscan, London.

Lockie, S., Lyons, K., Lawrencw, G. & Halpin. D. (2006) *Going Organic: Mobilizing Networks for Environmentally Responsible Food Production.* CAB International, Oxfordshire.

Lotter, D. (2003) Organic agriculture. *Journal of Sustainable Agriculture,* Vol. 21, pp. 59-128.

Mader, P., Flieβback, A., Dubois, D., Gunst, L., Fried, P. & Niggli, U. (2002) Soil fertility and biodiversity in organic farming. *Science,* Vol. 296, pp. 1694-1697.

Mason, R. (2011) North Sea tax impact 'marginal' says Huhne. *The Daily Telegraph* 5th May 2011

Maye, D. & Ilbery, B. (2011) *Changing geographies of food production.* In: Daniels, P., Sidaway, J., Shaw, D. & Bradshaw, M. (eds) Introduction to human geography. Pearson Educational, Harlow (forthcoming).

Michelsen, J. (2001) Recent development and political acceptance of organic farming in Europe. *Sociologia Ruralis,* Vol. pp. 41, 3-20.

Morgan, K. & Murdoch, J. (2000) Organic vs. conventional agriculture: knowledge, power and innovation in the food chain. *Geoforum,* Vol. 13, pp. 159-173.

Morris, C. & Potter, C. (1995) Recruiting the new conservationists: farmers' adoption of agri-environmental schemes in the UK. *Journal of Rural Studies,* Vol. 11, pp. 51-63.

O'Riordan, T. (1997) *Ecotaxation and the sustainability transition.* In: O'Riordan, T. (ed) *Ecotaxation.* Earthscan, London, pp. 7-20.

Plimer, I. (2009) *Heaven and Earth – Global Warming: The Missing Science.* Quartet Books, London.

Pocock, D. C. D. (1974) *The Nature of Environmental Perception.* University of Durham, Department of Geography, Occasional Publication.

Pulido, F. & Berthold, P. (2006) *Microevolutionary Response to Climatic Change.* In: Moller, A. P., W. Fiedler. & Berthold, P. (eds) Birds and Climate Change. Academic Press, London, pp. 151-183.

Redclift, M. (1987) *Sustainable development: exploring the contradictions.* Methuen, London.

Schahczenski, J. & Hill, H. (2009) *Agriculture, climate change and carbon sequestration,* National Sustainable Agricultural Information Service, www.attra.neat.org.

Scherr, S. & Sthapit, S. (2009) *Farming and land use to cool the planet.* In: Worldwatch Institute, 2009 State of the World: Into a Warming World, www.worldwatch.org/stateoftheworld.

Simmons, I. G. (1993) *Interpreting Nature.* Routledge, London.

Tovey, H. (1997) Food environmentalism and rural sociology: on the organic farming movement in Ireland. *Sociologia Ruralis,* Vol. 37, pp. 21-37.

United Nations (2010) *Report submitted by the Special Rapporteur on the right to Food, Oliver De Schutter.*

Walmsley, D. J. & Lewis, G. L. (1984) *Human Geography, Behavioural Approaches.* Longman, New York.

Willer, H. & Gillmor, D. (1992) Organic Farming in the Republic of Ireland. *Irish Geography,* Vol. 25, pp. 149-159.

Wilson, G. A. (1996) Farmer environmental attitudes and ESA participation. *Geoforum,* Vol. 27, pp. 115-131.

Wilson, G. A. (1997) Factors influencing farmer participation in the Environmentally Sensitive Areas scheme. *Journal of Environmental Management,* Vol. 50, pp. 67-93.

Wolpert, J. (1964) The decision process in spatial context. *Annals of the Association of American Geographers,* Vol. 5, pp. 537-538.

9

The Transformation to Organic: Insights from Practice Theory

Bernhard Freyer[1] and Jim Bingen[2]
[1]*Department of Sustainable Agricultural Systems, Division of Organic Farming, University of Natural Resources and Life Sciences (BOKU), Vienna*
[2]*Community, Agriculture, Recreation and Resource Studies, Michigan State University, East Lansing, Michigan*
[1]*Austria*
[2]*USA*

1. Introduction

This paper draws on practice theory to frame and understand the process of converting from non-organic (conventional) to organic farming. Within this context we seek to deepen our understanding of the transformation[1] processes that occur, including the on-farm experiences of farmers in the course of conversion to organic practices. More specifically, our aim is to

- Introduce general characteristics of the transformation process;
- Develop a theoretical framework based on practice theory which helps us understand the complexity of the transformation processes and
- Apply this framework in discussing selected aspects of transformation of an organic farm in the plant production sector;

We close with findings that are of theoretical and practical interest in understanding the transition to organic.

Practice theory offers a useful analytic means to identify and describe the essential or defining farms and related systems and dynamics of both non-organic and organic farming characteristics, as well as the related transformation processes from non-organic to organic systems. We look to practice theory for insights and understanding in the dynamic and reflexive inter-relationships between structures and individual performance, materiality and embodiment of practices and cognitive-mental processes.

Practice theory draws attention to the inter-relatedness of: the farmer's physical activity; the materiality of the things and artifacts with which the farmer works and which help to define the farmer's physical environment; and, the interactions between nature and the farmer as a social actor. We illustrate this process by looking at selected practices mainly in plant

[1] We recommend that the use of the term transformation in agri-food systems is used in a more broader, complex and holistic context going beyond the technical aspects of the farming system and to include linkages with systems outside of the farm (e.g. markets, input-industry, social networks etc.), while transition/conversion is mainly applied if the focus is only on technical and economic aspects e.g. of a farm.

production,[2] and try to specifically illustrate different patterns of social practices in all three phases of production from non-organic, through the transition and into organic.
From our perspective, the transition from non-organic to organic practices involves much more than following an approved set of directions or guidelines. For us, the transition involves a range of contradictory materialized and reflected worldviews. In this transformation farmers are challenged to rethink and reorganize their farming practices and their social networks. In order to capture the dynamics of this process, and to understand its agro-socio-cultural and political significance, we draw upon social practice theory to identify and generate new insights into the transformation process.

2. General characteristics of the transformationprocess

The conversion period from non-organic to organic farming practices is generally, and in some countries, legally, defined as a formalized process that is stipulated to occur over three years, during which farmers must follow all organic regulations (see Courville 2006; Greene and Kremen 2003; Greene and Dimitri 2003; United States Department of Agriculture 2002). On all farms, the transformation to organic involves a complex change in a farm's livelihood practices, including its internal and external (off-farm) social, ecological and economic relations.
Organic farming itself, and the transformation process on a farm, is discussed as an innovation process. More than that it is discussed as a step towards a paradigm shift in farming (e.g. Michelsen 2001, 3; Beus & Dunlop 1990). The rich literature about farmer's motives to convert to organic and how farmers convert their farms (e.g. Cranfield et al. 2010; Padel 2001; Fairweather 1999) explains, that this innovation process is not only initiated but is to be interpreted by general innovation characteristics identified from Rogers (1995), e.g. "relative advantage (more the better), relative complexity (less the better), compatibility (more the better), reliability (more the better), observability / trialability (more the better)." These mainly cognitive-mental arguments are of relevance, however they are different in organic and do not highlight the concrete practices, from which these reflections arise.
In contrast to other innovations in agriculture in the last decades, organic is not only a part of a new technique or a change of one practice on the farm, but a complete system change. Farmers adopt traditional techniques in a new context; conventional techniques are modified for the organic system; and specific new organic techniques are adopted. Besides these techniques, farmers follow specific rules and laws, which entail an ethical dimension. With this innovation they enter into new farm input systems, product markets and social systems, research needs and collaborations, and educational approaches.
The multi-layered process of transformation explains that the implementation of organic farming at the farm level does not follow a linear process or an S-curve of diffusion processes (Rogers 1995; Ryan & Gross 1943) over time. The diffusion of this innovation on a farm seems much more a back-and-forth process, which is influenced by the farm's internal and external changes. It follows loops, path dependencies (Latacz-Lohmann et al. 2001), and surprising developments. It is a contingent process of adoption and adaptation of old, modified and new practices which have to stand the test of time. This transformation process is embedded in and shaped by a specific space-time, cultural, social, economic and an agro-ecological context. The transformation process is formally finished after three years.

[2] We are aware that the specific focus on plant production might exclude relevant aspects to fully interpret the phenomena of transformation.

However, the adoption and diffusion of practices is an ongoing process, and each sector of the farm has its own specific processes of change and innovations. Therefore, we argue that the transformation process is a non-linear cyclic innovation process (cf. Cheng & Van de Ven 1996). The transformation process is not a predetermined process and it neither follows one pattern nor ends up with the same result. On the contrary, there are differences concerning the extent of "change"; e.g. between farm types (e.g. dairy farms, vegetable or hog farms), agro-ecological zones (e.g. mountainous or arable regions), cultural impacts (e.g. farmer-consumer co-operations or export oriented production; country and region specifications) and farmers following a resource limited approach and those who are engaged in "substitution organics" (see Guthman 2004a; Guthman 2004b), These different types could be classified as transformation types, defined by their starting point and their target intensity and the degree of change. However, types presented in Table 1, are not more than a simplified picture of this diversity.

Transformation type / farm level	**Low change**	**Low change**	**Low-medium change**	**a) high b) low change**
Starting intensity non-organic	Low input*, part-time farmer	Low input, fulltime farmer	Integrated farming, fulltime farmer	High input, part-time or fulltime farmer
Organic target intensity (Agricultural intensity)	Low input**, part time or fulltime farmer / often modernized	Low input, fulltime farmer / often modernized	Low-medium input, fulltime farmer / modernized	a) low input b) high input / fulltime farmer / modernized

* Input: Herbicides, pesticides, mineral fertilizers, livestock unit per ha

** In the framework of the Basic Standards: mineral (Phosphorous, Potassium) and organic fertilizers and "organic" certified pesticides; livestock unit per ha; share of fodder legumes in the crop rotation

Table 1. Diversity of transformation types

A wide variety of issues related to the transformation process have been studied over the last two decades. The driving forces for transformation to organic farming involve a broad set of motivations, which might be environmentally, economically, religiously or ethically driven (Cranfield et al. 2010; Khaledi 2007; Locke 2006; Engel 2006; Darnhofer 2006; Darnhofer et al. 2005). According to Cranfield et al. (2010), we differentiate between four types of farmer motivations, which are approximations of what we find in reality:

- To find an economic solution for the farm: economic survival, market strategies, farm reorganization
- To take care for the environment: nature protection, water protection, soil fertility
- To avoid risks and to increase health: to exclude unhealthy methods, to recover health
- To follow and fulfill idealistic motives: to live a self-realized, spiritual, religious, value driven life

These different motivation types underlie the transformation process and go beyond a change and reorganization of techniques. This observation also explains that there are more than enough reasons that the perspectives of researchers on the transformation processes highlight a broad spectrum of topics. Some of these are: the challenges in the transformation period in a broader context (Lamine & Bellon 2008; Padel 2001; Tress 2001; Lockeretz 1995; Freyer et al. 1994; Lampkin 1994; Freyer 1991; Rantzau et al. 1990); production and economic (Schneeberger et al. 2002; Dabbert 1994); investment (Odening et al. 2004); market processes (Tranter et al. 2009); transformation planning (Goswami & Ali 2011; Ács 2006; Freyer 1994; Lampkin 1992; Dabbert 1991; MacRae et al. 1989); and the systemic characteristics of the organic farming (Darnhofer et al. 2010; Noe & Alroe 2006, 2003; Høgh-Jensen 1998). Despite their different theoretical perspectives, all of these studies sensitize us to the multi-faced characteristics of the transformation process.

Drawing on this wealth of empirical and conceptual discussions of organic farming, as well as our own experiences about the complexity of transformation processes at the farm level (Freyer 1998; Freyer et al. 1994; Freyer 1994; Freyer 1991; Rantzau et al. 1990), regional transformation processes (Freyer et al. 2005; Freyer et al. 2002; Freyer & Lindenthal 2002), participatory approaches (Wehinger et al. 2002) and our initial efforts to apply practice theory (Freyer et al. 2011), we argue that there is a need for new theoretical perspectives on transformation processes.

3. Conceptualization of practice theory

Before we describe our model, we first explain more specifically our interest in practice theory and offer some insights into this theoretical perspective. From our own life experiences we know what the challenges of changes in habits can be. Change occurs in thinking, feeling and in what we communicate. However, change becomes most visible when we engage in a practice and when this activity becomes part of a routine in our life. Similarly, we suggest that rethinking the transformation from non-organic to organic farming requires the use of a theoretical perspective that captures the practices of farm transformation from a non-organic towards an organic farming system.

Theories with a broader perspective for analyzing complex systems change are systems theory (Bertalanffy 1973), social practice theory (e.g. Schatzki 1996) and transformation theory (e.g. Reissig 2009). Social practice theories play an intermediate and integrative role between systems and transformation theory. We argue that system theory, which is sensitive to actors and actants, offers a methodological feature that draws attention to the pre-structuring of relevant factors, - in our case the material - which is part of the social practice. In addition, it also highlights the ways in which humans participate in the practice as well as in the exchange of materiality. However, it underestimates the individual dimensions such as cognitive and mental processes or how individuals act.[3] Transformation theory focuses on macro level development processes, on organizational and societal structural change (Reissig 2009), but it is not sensitive towards the individual, materiality or embodiment of practices. Therefore we redefine and extend the term transformation with reference to systems theory as well as practice theory and to processes at micro-, meso- and macro-level.

So, why practice theory? In the transformation process from non-organic to organic farming the farmer specifically confronts changes in those practices that were learned, created, sustained, and part of social experience and reproduction in everyday life. The transformation requires changes to embodied practices that made sense for a long time, and that informed what was always done, and was confirmed by the cultural and social context (Schatzki 2002).

To describe farm and farmers transformation, we seek a theoretical concept that is sensitive to the farmers' individual perspectives, the changing material, social, knowledge, mental and structural conditions confronted in this period, and that guides or influences decisions. Practice theory offers such a broader view on human behavioral change. It is sensitive to those aspects, which are fundamental in change processes, e.g. (Strengers 2010, 17): "How are everyday practices reproduced in daily life and what, if anything, disrupts these

[3] This observation however will be deepened in a further paper.

routines? What rules constitute specific practices and how do they affect what individuals' do? To what extent do individuals past experiences and upbringings influence their practices? How and why are practices changing?"

In the last two decades various analytical approaches were developed which could be called "Theories of Social Practices" (Reckwitz 2003, 282). Practice theory seeks to overcome the dichotomy between structure and action, and to understand structures as the repetition of micro-situations (Collins 2000, 107) as well as the habitus of people as a product of social conditions (Bourdieu 2005, 45). Furthermore, practices are built on materiality of things and artifacts in a specific arrangement, which relate to a space-time context, follow routines, but are also open to change. A series / bundle of these social practices create lifestyles, constitute structures or social fields - networks, organizations or institutions (Reckwitz 2003, 285, 294, 295). These routines of behavior, which emerge as interplay between actor and actant, together create the site of the social (Reckwitz 2003, 287), in contrast to most cultural theories in which the site of the social is a cognitive and mental-intentional structure (Reckwitz 2004, 318). Practice theory is not a cognitive scheme, or something embedded in discourses and communications, but a practical knowledge, a know- how, a series of every day life concrete practices (Reckwitz 2003, 287) which are reproduced, routine-embodied performances; "objects are handled, subjects are treated, things are described and the world is understood " (Reckwitz 2002b, 250). The understanding of this practical knowledge includes both - "consciously reflected and semi- or deeply embedded knowledge" (Strengers 2010, 8).

We argue that there is need to add cognitive-mental and structural perspectives with the concrete practices and the related materiality. There are several reasons to adopt this perspective. As Nicolini (2009) summarizes:

> "the meaningful, purposive and consistent nature of human conduct descends from participating in social practices and not from the deployment of rules, goals and beliefs" (Nicolini 2009, 4); Further more: „practices constitute the horizon within which all discursive and material actions are made possible and acquire meaning; that practices are inherently contingent, materially mediated, and that practice cannot be understood without reference to a specific place, time, and concrete historical context" (Engeström, 2000; Latour, 2005; Schatzki, 2002; 2005; cit in Nicolini 2009). While practices depend on reflexive human carriers to be accomplished and perpetuated, human agential capability always results from taking part in one or more socio-material practices (Reckwitz, 2002b; cit in Nicolini 2009,5). Practices are mutually connected and constitute a nexus, texture, field, or network (Giddens, 1984; Schatzki, 2002; 2005; Latour, 2005 Czarniawska, 2007). Social co- existence is in this sense rooted in the field of practice, both established by it and establishing it. At the same time, practices and their association perform different and unequal social and material positions, so that to study practice is also the study of power in the making (Ortner, 1984; cit. in Nicolini 2009, 5).

Transformation is not only a change of one technique, it's a systems change that involves social relations and structures; it is a far reaching break with former practices (Reckwitz 2002b, 255), common understandings, how to do things, following certain norms, conventions, customs, traditions, and what is acceptable in practice or not (cf. Turner 1991). If we follow Bourdieu (2005, 47) the characteristic of habitus change is constantly and continuously a change between historical given structures and new practices. With the transformation process, the farmer moves a big step forward and rejects most former social practices.

To conclude, we picture our approach to organic transformation as an interplay of four analytical-theoretical dimensions: structure, individual performance, materiality and embodiment, and cognitive - mental processes (see Figure 1). In the following we describe our understanding of these dimensions, their relation to social practices and their inter-relatedness.[4]

3.1 Structures

To frame the dimension of structures we draw first from Anthony Giddens' approach to structuration. Giddens (1992, 1984) identifies structures (orders of knowledge) as sets of rules and resources (see Westermayer 2007, 10) (Table 2).[5] Structures are interpreted as rules with a regulative and constitutive dimension in a space-time context and as material and power resources. Rules and resources are established in and through practices, by doings and sayings.

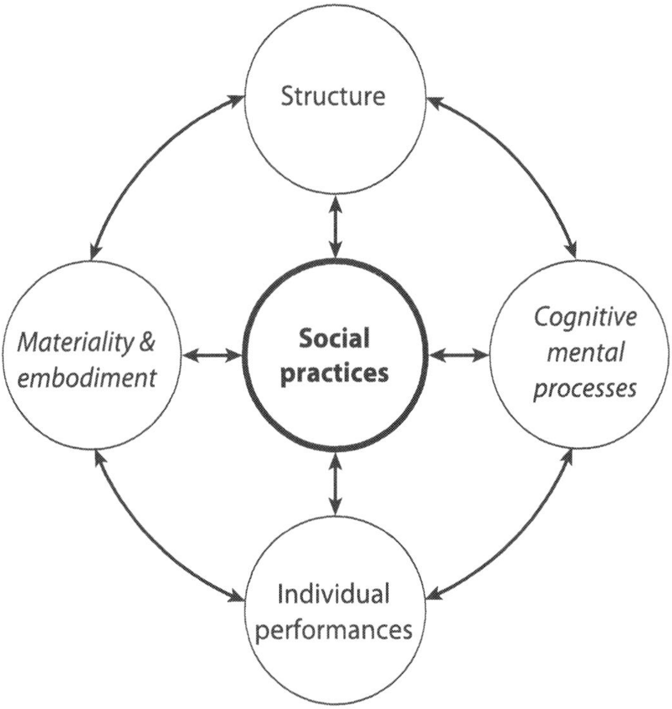

Note: All dimensions are linked and they influence and form each other. They are stored in structures and arise through carrying out social practices, defined as a bundle of practice arrangements, which exist independent from individuals. Individuals take over but also perform the social practice beyond their own materiality and embodiment and as well cognitive-mental patterns. Social practices combine objective behavior and subjective sense.

Fig. 1. Framing A Theory of Social Practices

[4] For other models on practice theory see e.g. Brand 2011

[5] Anthony Giddens understanding of structures contrasts with that of Ralf Dahrendorf's (1973) theory of conflict sociology and Talcott Parsons functional part systems of society (AGIL scheme)(1970)

Rules stipulate how to engage different resources, how they are socially shared yet remain anchored individually (Giddens 1992, 316). Two different types of rules can be identified: those that are constitutive, or produce sense and significance, and are communicable; and, those that are regulative because they limit or encourage action. Norms are reflected and embedded in cognitive structures of knowledge (Reckwitz 2004, 315). With practice theory the character of rules and norms is also a practical one. The reproduction of the social practice (Giddens 1984, 21) involves a cognitive, conscious reflexivity (Collins 2000, 107); "rules also emerge out of practices, and are often interpreted and incorporated into practices in ways different from those originally intended" (Strengers 2010, 12, modified). Resources can allocate, or have the power to change something, or they can be authoritative, or have the ability to influence the acting of others.

Orders of knowledge are formed by social practices. In turn, they reproduce, reformulate and modify practices, and empower or hinder acting. Orders of knowledge do not exist without a social practice. They become relevant in the execution of a social practice and these practices reproduce institutional orders (Giddens 1984). In the meantime, they support a sense making identification and selection of social practices.

Structures are formed by materiality (e.g. things, artifacts), and they are also initiated through cognitive-mental processes, which form orders of knowledge through discourses, norms, rules or laws (see below). However, and this is the specific characteristic of practice theory, structures always exist only within a context to any practice.

			Medium of acting	**Institutional order**
Structures	Rules (shared knowledge)	Constitutive aspects	Foundation of meaning, communication	World views, discourses, codes
		Regulative aspects	Moral relation, sanctions	Moral, legitimate order, laws, norms
	Resources (tools)	Availability of allocate resources	Material power of arrangement	Economic institutions
		Availability of authoritative resources	Power over other actors	Political institutions

Source: Westermayer (2007, 10, modified); Giddens (1984)

Table 2. Model of Acting based on Structuration

3.2 Materiality and embodiment

The concept of materiality and embodiment draws our attention to the materialization of the social and the cultural in objects, the embodiment of practices, and the engagement with things and technology and in our context also specifically with natural resource management (Pali et al. 2011). In practice theory, things (e.g. artifacts, bodies or natural objects) are re-conceptualized as social entities that contribute to the formation of practices (Preda 1999, 349).

As Schatzki (1996) argues, materiality represents a constitutive element or resource of social practices. To consider materiality conceptually means to consider natural objects and as well as man-made artifacts, including characteristics e.g. smell, taste, odor, sound, form, structure or function.

Following Reckwitz (2003) and others, we apply the idea of embodiment to define collective social practices by drawing attention to the site where practices are embodied and expressed by certain physicality. Things and embodied practices are also the carrier of rules or regulations, resources or certain knowledge (Reckwitz 2002a). Things represent rules and norms but they are also the source for their modification or rejection.

Material infrastructures are also of high importance in our investigation, because some investments in farms can be extremely difficult to change. Many are long lasting and path dependent, e.g. hog houses. Some hinder change, depend on their amortization costs (see also Arthur 1989). This path dependency could have a huge impact on change and with that on practices (Latacz-Lohmann et al. 2001).

3.3 Cognitive-Mental processes

Cognitive (perception, cognition) emotional or affective processes are "non-material" processes, …" as well as cognitive bases of behaviour" (Warde 2006, 140). Cognitive-mental processes are part of individual and group practices, and they contribute to the formation of structures. Rules for example are based on, and integrate cognitive processes (see teleo-affective processes below). Knowledge processes arise in discourses, they contribute to orders of knowledge and they are part of a practice.[6] Discourses are specific social practices of representation, in which cultural codes are manifested (Reckwitz 2006, 43, cit. in Jonas 2009, 10). Codes are part of these social practices and enable these practices (Reckwitz 2008c, 17). Culturally formed codes transport the sense, they differentiate between 'in - and outside', that what is part of a system and that what is excluded (Jonas 2009, 10). "These conceptualized 'mental' activities of understanding, knowing how and desiring are necessary elements and qualities of a practice in which the single individual participates, not qualities of the individual" (Reckwitz 2002b, 249, 250).

3.4 Individual performance

Individuals perform, or they adapt and adopt their practices to structural realities on their own terms (Greve et al. 2009). Practice(s) is created in the mind, but subject to the availability of things and artifacts, as well as an individual's capacity and capability to embody practices (Reckwitz 2006, 40); they are always related to an individual's concrete practice. The adoption of practices by an individual is also heavily influenced through the socialization process of an individual (Shove & Pantzar 2005, 2007, cit in Jaeger-Erben 2010, 254), e.g. former execution of specific social practices and therefore generalization of these practices seems limited.

Through the social practices, an individual creates her/his own social position (Reckwitz 2006, 34, cit. in Jonas 2009, 13) and an identity. This understanding follows Bourdieu's habitus concept (Pouliot 2008, 273, 274), with respect to these characteristics: interpretation from a historical perspective, formed through individual and collective experiences; a reservoir of practical knowledge, learned through practicing in the world; relationality learned through inter-subjective experiences; dispositional, in the sense that that habitus does not determine forms of action in a mechanistic way, but encourages actors towards specific actions.

For Reckwitz (2002a, 207), the individual's subjectivity is created through cultural codes, and individual capacity for reflection (2006, 40). That is, individuals establish their own

[6] See also Reckwitz's discourse / practice approach (2002b, 2008b)

individual practical understanding; they are partly autonomous actors (see Rammert 2008; Thevenot 2002, 69). They have the ability to give meaning to their thinking and action (Jonas 2009, 18). The meaning of a practice is given by the practice itself (practice specific) and by the actor (actor specific sense), however, the practice itself exists independent from the actor. From this perspective, practical understanding does not mean that a rational actor holds a normative meaning to action, but steers the action by "conferring meaning" on it (Jonas 2009, 3).

3.5 Social practices

Based on the above-introduced dimensions, which are from and contribute to social practices, in this section we synthesize and re-conceptualize this theory from the perspective of social practices and their key characteristics.

First, a social practice can be described as a routinized, and physical performance (see Rasche and Chia 2009), or a spatially dispersed nexus or pattern of physical activities and observations on these activities (Schatzki 1996, 89). Social practices are "bodily-mental routines" (Reckwitz, 2002b, 256). These practices, or bundles of practices, are not homogeneous, but full of contradictions. From the perspective of systems theory, we talk about open, complex systems (Berkes et al. 2003) that are in a permanent process of reformulation, while temporarily static (routined habits). Practices exist independent of individuals, and constitute rules and resources. Social practices lead to material consequences and embodiment. There is always a cognitive-mental dimension 'participating' in these practices, but the majority of the practice theorists agree that social practices assume a leading position.

Second, practices are "ordered across space and time" (Giddens 1984, 2). They produce new social space through practices, as individuals become member of this space. They assume and practice rules and resources (e.g., any type of power) that exist independent of each individual (Bourdieu 1989). The site in which actors "perform" prefigures the acting. But actors and actants create a network of orders and practices (Schatzki 2002, 63) that is related to specific characteristics of sites. Practice also entails a specific conceptualization of time in which it is structured; and, each actor carries an individual interpretation and practice over time.

Third, practices are contextualized in social fields. Social fields are more or less differentiated and institutionalized concepts of complexes and networks of social practices and cultural discourses (Reckwitz 2004). A field is a common space of knowledge in which specific practices are legitimated or not; it is a "playground" in which actors become socialized to certain rules and resources and in which actors assume positions and try to optimize their social resources (Bourdieu 1993). Any practice and any thing also embody morality or immorality (Jelsma 2006, 222), i.e., fixed in rules, laws or regulations or culturally embedded as a common sense.[7]

Fourth, practices can be characterized as having different features. Those which arise in different contexts, but which always are more or less in the same form, are described as *dispersed* (following rules, imagining, describing) (Schatzki et al. 1996, 98). They contrast

[7] For example in organic agriculture the threat of nature through pesticides is interpreted as immoral, something that is excluded by their regulations. Acting with pesticides is an embodied social practice in non-organic farms and for those farmers under certain circumstances not immoral. In contrast, it "protects" against a threat.

with *integrative* practices, which cover complex entities. There the acting and speaking is diverse, but is combined through practical understanding or a set of rules and a teleo-affective structure (goal oriented; combined with emotions etc.). Practical understanding is thus the capability to do or say something.

Fifth, some practices are in a certain sense self-evident (non-discursive), while others are discussed (i.e., discursive) (Reckwitz 2008a). Non-discursive practices are socially normed and routinized forms of behavior, which comprise certain knowledge, or "know how", including interpretations, motivations and emotions. Discursive practices on the other hand are temporary and intended to be the starting point for new practices (Schatzki 2002, 85). Reckwitz assigns the discourses as primary generators of meaning for orders of thinking and saying (2008b, 193).

Sixth, an additional approach for distinguishing practices is how actor and actants are involved in their practices. Inter-subjective practices ask for more than one actor and are discourse oriented. Inter-objective practices are those between objects or artifacts and are self referential, if the practice is directed towards the individual (Reckwitz 2002a, 206). Adler (2005) describes in this context, communities of practices, as communities of common distributed actions instead of a common organizational structure.

4. The social practice perspective of agricultural transformation processes

4.1 Approach

In order to understand this overall process of transformation, it is important to have a broader view of the farm history from the non-organic period through the start of acquiring organic status. For the purposes of discussion we compare a 'typical' high-intensive non-organic farm with an 'ideal type' of diversified organic farm[8]. In addition, we use the case of a farmer who wishes to become certified, according a set of national organic standards or regulations. Finally, we consider the transformation process as one of moving from non-organic farming practices through a series of practices to becoming 'organic'. We use plant production with a focus on selected practices, such as crop rotation, weeding techniques, pesticide and fertilizer use in the three phases - non-organic, organic and the transformation towards organic.

The diverse farm histories, the complexity of agricultural practices, the diversity of farm structures and agro-ecological realities, farmers and their families and the understanding of a farm as an individuality, all of these teach us to be cautious with generalizations of farming arrangements and practices. However, there are some patterns, which characterize the types of farms on which we will focus in the following analysis. These are the arrangements -"assemblages of material objects, persons, artifacts, organisms, and things" - (Schatzki 2006, 1864), which are widely accepted as part of farming practices and the structural and organizational environment of the farm. However, these only partly inform how the practices are carried out.

We select several relevant characteristics, which frame non-organic and organic farming (Figure 2) and which describe the main artifacts and natural objects, etc. used in plant production (Table 3), and to make the systemic character of agricultural approaches more explicit. Moreover, we introduce with Table 3 the paradigms (cf. Lorand 1996, modified; Beus and Dunlap 1990; Guba 1990) of non-organic and organic agriculture in order to serve

[8] We do not consider the multiple changes or practices after a farm is recognition as an organic farm.

as the underlying, model or ideal-type, orientations in acting and decision-making processes in the non-organic and organic world.

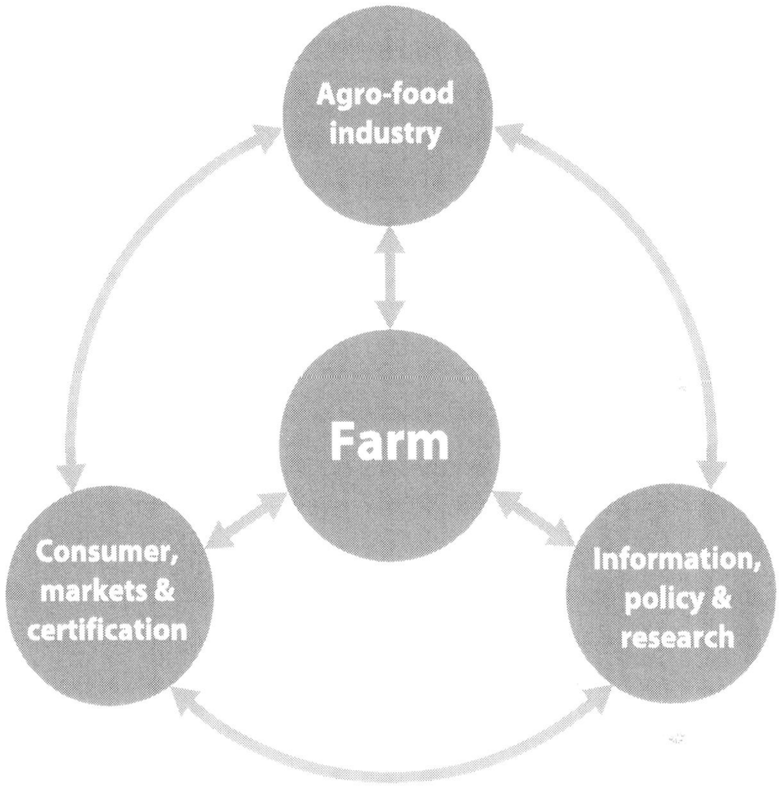

Fig. 2. Farm related environment

4.2 Non-organic farming

The non-organic corn-soybean system is largely possible on flat arable lands that enable a type of farm style and size associated largely with the Midwestern states in the US. It creates a linear landscape structure of very large, "fencerow-to-fencerow" fields[9] designed to facilitate the use of very large equipment. The practice of corn-soybean cropping creates and organizes a particular type of landscape, based on specific arrangements of artifacts. This type structure of agro-industrial landscape dominated by extremely large machinery and absent of people illustrates what Schatzki refers to as the co-constitutive relation between practices and arrangements (2010, 140).

[9] Field: agronomic unit; social field as part of a social space (Bourdieu 1990, 1988, 1983): the entity of societal interactions and constellations; fields are e.g. politics, economy or arts: sub fields are e.g. field of literature, of school or university, which differentiate in the field of literacy and the cultural field

Characteristics	Non-organic	Organic
	Farm	
Plant production and agro-ecology	High input system (mineral fertilizer, herbicides, pesticides, GMO, technology); mono-cropping, large farms, large field size, low amount of biotopes; Segregation between farming and nature protection	Low input system (limited fertilizer input, mechanical weed control, mainly biological pest control); crop rotation, legumes, biotope rich; compost, green manure; integration of nature protection into the farming system
Animal husbandry	High performance, large animal groups, high fodder input (concentrates), short life span of animals, antibiotics, hormones allowed, slurry production, confinement	Low input, mainly own fodder, low amount of concentrates, long animal life span, antibiotics regulated, hormones excluded, often stable manure, ethical rules; pasture and farm yard as run-off, free-range
Farm economy and market orientation;	Industry oriented, controlled by industry; one market, economy of scale	Diversified production, investment into soils and biodiversity, several markets
	Related environment	
Agro and food industry	Global players, uniformed commodities	Local and regional players, diverse, partly international
Consumer and markets; certification	No contact with the consumers; mainly big retailers; voluntary certification systems	Diverse relations towards consumers and different markets; certification following specific guidelines, also linked with subsidies
Information; policy (subsidies); research	Industry; compatible with official agricultural policy, sector (commodity) oriented subsidies; research mainly by private companies and private financed Universities	State and farmer organizations; partly in line with policies, environmental friendly oriented subsidies; farmer driven research, University research mainly state financed
	Paradigms	
Ontology	Short term profit oriented; maximization of labor and technological efficiencies	Farm is part of a broader eco- and socio-cultural system; natural conditions are accepted and adjusted
Methodology	Technology and output oriented	Balancing between the different parts of the farming system, avoiding losses
Epistemology	Observation, analysis and policy decisions, technological framework	Observation, diagnosis, and therapy, prevention and risk avoidance

Table 3. Characteristics of Different Farm Approaches

A truck delivers the seeds directly to the farm or the farmer has to order from the regional seed company. With the seeds, the farmer receives a complete set of inputs, which includes mineral fertilizers, pesticides and herbicides. These artifacts are the preconditions to organize and to conduct this type of farming. The practice to order these agricultural inputs does not require a long search process; all is advertised and supplied by firms; purchasing these inputs is a routined process of given bio-chemical materialities. These practices take place every fall and spring, at the same sites with the same technology and within the same social web; they are site and time specific.

The practice of this style of farming needs a certain materiality of big tractors with an air conditioned cab, driving up and down a field with a size of several football fields, in a cleared out landscape, plowing the soil, talking on a cell phone, watching a movie or TV. It means sitting in a tractor, and activating electronically controlled steering mechanisms. The physical landscape, the technology, but also the knowledge order given by agro-industry, the farmer translates into a certain agricultural practice in his specific farm context; they all hang together and the practice would not exist without each of these artifacts. These practices became routines and are unquestioned orientation for practicing corn-soybean agriculture. They are also confirmed in official statements, documents, internet pages, or news journals of the agricultural administration and industry and by a majority of agricultural researchers.

The farmer handles the tractor computer with a GPS to spray the exact amount of mineral fertilizer or pesticide, using a technique prepared and recommended from agro-industry. The farmer (or any company using airplanes for spraying) repeats the pattern of practice carried out while plowing: driving precisely row by row, up and down the mainly rectangular field. In the tractor cabin and "protected" from outside, the farmer is unable to listen to or smell nature.

The site where the inputs of farming are created/produced/prepared/collected is off the farm. At the same time, "farming practices' are carried out by one person in "cooperation" with technological artifacts. Interaction with nature is limited as electronic devices, which "interpret" nature and guide the practices, give most of the relevant information for the practice. The moral justification for using chemicals exists in the conviction of 'we are feeding the world'.

The industrial artifacts with bio-chemical and biological characteristics are transported over long distances, figured out by humans who are acting in an anonymous web of relations, and materialized in communication technologies. Machinery (tractor...), the biochemical and biological artifacts and related practices to produce them, are dislocated from the farm over long distances. They move towards farms and are stored temporarily in the farm barn. For that, the farmer uses a front loader to stack the bags in one part of his barn. These practices describe parts of the energy flow, and the accompanying technological and social constellations, which enable the transfer/movement of materiality from site to site - from outside to inside the farm, and from site to site within the farm.

Harvest might be done together with additional personnel or is out-sourced to specialized companies. Efficient use of harvest technology again requires a certain field size and form - landscape design is also part of the specific harvest practice (cf. Shove et al. 2007, 134). Harvest itself takes place in several days, again with powerful technical equipment. Parts of the harvest might be stored on the farm in a storehouse, which do not look any different from buildings in a suburban industrial zone, or is directly transferred with large trucks towards a train station or a regional storage building. And finally, the truck dimensions require a certain size of road system in the landscape between the fields.

Farm practices require certain architecture, and architecture enables the practice. Using inputs from outside the farm has its impact on farm architecture, barn structure, storage buildings and daily routines, the design of the farm yard and the stretches of roads and paths in the farm yard, as well as towards the fields or to buying and selling points for agricultural inputs or the harvest.

The materiality of non-organic agriculture pre-structures the daily practices and the labor distribution year round, it forms the landscapes, the colors of nature, the smell and the structures of the fields, the soil and water quality, and the (non-) existence of biotopes or the biodiversity of plant and animal species.

Fieldwork constructs a farmer's annual schedule. The farmer waits for best weather conditions, or is informed by the regional agriculture weather report, when to start with fieldwork. Fieldwork is done in two, or a maximum of three time periods in early and later spring and finally at a crop's specific harvest times by the farmer or with workers. These industrial and nature preformed practices not only structure daily life, but also occur at specific times over the year and thereby structure the whole year on the farm, largely independent of any individual decisions.

Farmers slightly modify their practices every year by adopting new industrial products and adapting to changing weather conditions. Tolerance for reformulating or reshaping the practices by the farmer is limited. Natural science and technological knowledge is embedded in the artifacts, produced and recommended by a disciplinary oriented science. Industrialized agriculture is an example for social practices which are characterized by reproduction, while change is limited (Reckwitz 2002a, 255, in Bueger and Gadinger 2010, 289). Farmers are the carriers of the materialized industrial products on the landscape, and landscapes in turn are patterns of these practices.

The materialized production program also influences relations with consumers - the dimension of production excludes direct marketing in a region, and with that, it limits the social network potentials. This type of agri-culture does not communicate with a region, because its products are unable to play this role; instead, it creates a monotone landscape whose message is: "stay out". As such, production does not "produce" social relations or a local culture. Many successful conventional farmers take over more fields from the neighbors; essentially, the farmer is "buying out" his community. The agricultural related social network is mainly limited to the industrial partners.[10] As players in the global network, the relation with the farmers is that of a business relation.

In summary, the social practices of non-organic agriculture are often and specifically in this described case, dominated by the materiality of industrial and industrialized inputs, embodied and habitualized in concrete farming practices as well in the structure of the farming landscape. These practices became routines and serve as unquestioned orientation for practicing corn-soybean agriculture, also confirmed by the dominating (agro-industrial) media. While the production risks are in principle assumed by the farmer, significant public subsidies to maintain industrial farming help offset these risks. Practicing the same agriculture, the farming practices of the farmer are accepted by their neighbors. Daily information by Internet and glossy brochures by industry guide the practices. Preparing the

[10] From Schatzki (2010, 128), the site is not a geographical but a social dimension, ..." pertaining to human coexistence, ..." the hanging together of social life's".

annual crop business, the farmer studies in the Internet the global stock market, contacts the dealer electronically or by cell phone.

In the socialization of industrialized agriculture, these practices have been for decades non-discursive. Those who follow this practice are on the one hand path dependent, and on the other hand supported by the agriculture and food industry (inclusive consumers), research and education system - all reinforcing this type of farming, partly criticised, but mainly an accepted habit in current Western societies.

4.3 Organic farming

What are the defining practices of an organic farming system? All organic farms (except for some large-scale industrial operations, are organized around highly diversified cropping systems involving more than five different crops organized in a crop rotation and inter-cropping systems. In this type of farming, each crop plays a different role in mobilizing nutrients, creating soil fertility, weed-control and pest and disease management. Instead of mineral fertilizer from outside the farm, organic manure management techniques are essential to the system of practices.

One scenario of the inter-related and reinforcing practices can be outlined as follows: the diversity in crop rotation affects the weed diversity. Plant protection and soil fertility management in turn is related to cropping practices designed to manage long-term soil fertility (see Freyer et al. 2011). Diversified crop rotations are essential in pest and disease management. Fields are usually quite diverse in size and structured with a multitude of crops that play multiple roles in the overall farming system, including the maintenance of different habitats such as hedges and grass stripes that play a range of roles from soil to wildlife management. Equally important, no use of synthetic chemicals helps to foster soil health and diversity; "healthy soils mean healthy plants." In short, there is a systemic and systematic set of diversified material interrelationships built into and generated from these practices. They are all of a specific material quality and require a specific technology in handling things, artifacts and embodying organic objects.

While crop rotation is part of non-organic farming, it is a pivotal element in organic practices. Farmers must have an intimate knowledge of their fields, including field histories (how they "behave" under different conditions over time) that help identify the ways in which previous cropping practices could influence current cropping practice, such as planting legumes to fix nitrogen. That is, in an organic system, all the agricultural practices account for time, both the past and expected future of a field, including each field's relationships with other fields through organic material transfer or the habitat function for both pests and beneficials, which move from one field to the other.

In a similar way, plant protection or weed control is not first a matter of using industrial products. It involves: the arrangement of crops over time and space; the specification of the tillage system following the specific demand of crops; the biotope diversity in structure and function and distribution in the landscape; and the history of farming practices and the current qualities of a soil in a specific field. The farmer is at the center of this process, bringing old and new perceptions and understandings and an ever evolving discovery and understanding of organic practices. Organic farming becomes a self- referential rethinking of practice and inter-objective relations with nature. In many cases, organic farmer's neighbors who often practice non-organic, offer little or no advice or relevant experience from which to learn.

Diversity in cropping involves diversity in the material arrangements for each crop (weed control, addition of organic manure, etc.); sowing and harvesting dates differ and lead to diverse cropping activities throughout the year as well as a diversity of relationships with "nature," or uncultivated spaces. Intimately aligned with this, organic farmers seek to create the most favorable living conditions for soil organisms and plants by relying on natural processes and largely without external (to the farm) and synthetic inputs. This means that the source of the materiality contributing to plant production is located on the farm, and produced on the farm. The farm site, and not external, much less industrial actors, becomes the farm's principal resource. Equally important, each organic practice, while it may be performed at a specific moment, occurs over time and is often cumulative. The site-specific agro-ecology of each farm reinforces the overwhelming importance for each organic farmer to adapt and act on a specific site - the whole farm including the relationship among different parts and sites on the farm as well as those off and around the farm.

The dominating organic materiality leads also to different forms of embodiment, which includes more direct contact to nature. The exclusion of synthetic chemicals generates a far-reaching chain of activities, including no purchasing, storing, processing and spraying. Instead, farmers become involved with the management of organic manure by establishing compost heaps, their coverage, and for some, hand mixing of bio-dynamic preparations in a barrel following rhythms for a specified time.

Farmer's decisions regarding practices involve reflexive engagement. This is necessary because organic arrangements represent a redistribution of materiality, actor-actant relations, and the distribution of materiality, arrangements and practices over space and time. Diversity, limited inputs from industry, the dominant role of "organic/natural" objects and techniques, which support the self-regulation of nature, are defining characteristics organic farming plant production practices.

The limited impact of standardized artifacts and the broader contribution of natural objects and the complex, unforeseen nature-based processes require more interpretational activity by the individual; therefore to follow routines or concepts of organic practices has its limitations. The farmer's decisions about carrying out practices require reflexive engagement.

Contrary to the conventional practices, it's not first of all the industrial pre-formed technology, which shapes the organic practices, but the natural given objects, rules and ethical principles. Diverse farm patterns and the indeterminacy of the production process, given by complex natural processes, explain the learning processes.

There is a discursive character to the social practices in the organic context. Beyond the materiality, organic practices commonly overlap with interests of diverse groups of non-farming actors, often members of different social fields (e.g. artist, environmentalists). Practicing organic is communicative and discursive; this discursive character results from the complexity of the cropping systems with diverse practices over space and time. Consequently, the individual performances and engagement varies widely and involves a wide spectrum commonly individualized farming activities.

The limited involvement with industrial interests also influences the political and policy relationships in organic farming. The power, and the freedom of the farmer, is defined by: the state regulations that are often a precondition for receiving financial subsidies; the certifying organization that stipulates the rules to follow and the sanctions for non-compliance; and, the "voice of the market" including local consumers who come to the farm or a farmers market who prefer relationships built on trust and direct connections with the producer over those in supermarkets.

General conceptual "knowledge of orders" is formed by discourses on environmental and ethical standards, introduced by the IFOAM Principles (2009) and basic standards for organic farming. The arrangements of artifacts and the practices represent the transfer of the IFOAM principles (health, ecology, fairness and care) and basic standards into social practices, which include the ethical discourse and worldviews of the organic movement. Schatzki (2005, 480) describes this process as follows: "Understandings, rules, ends, and tasks are incorporated into participants' minds via their 'mental states'; understandings, for instance, become individual know-how, rules become objects of belief, and ends become objects of desire". Principles serve partly as a "must"; norms, regulated by law, serve partly as recommendations. They represent the cognitive basis of organic farming, which regulates the practices, and influences the handling of soils and plants. Principles include the human-nature-relation, based on a deep respect for, and effort to understand nature that includes the exclusion of chemicals from the production. They represent a result of discourses based on farmer's long-term observations, practices and research. Their implementation is discursive, and site-specific agro-ecological conditions, asks for a reflexive use, reinterpretation and adaptation.

Organic creates and relies upon the creation of social networks (e.g. Jarosz 2000), on and off the farm, and thereby creates communities of a culture. Consumers communicate with farmers about products or their worldviews. In other words, each "farmer - consumer community" involves a practice of sharing time, doings and sayings, which occur in specified social fields and take place in specific sites (cf. Hörning 2001, 160).

4.4 Transformation towards organic farming

It is simple, but important to understand that the transformation of the farm involves a move/change from well-understood and long-practiced non-organic activities into a series of unknown organic and unknown transformation-specific practices. As Schatzki (2006, 1068, 1864) reminds us, the actions constituting a farm history include a practical understanding of non-organic farming practices, a well-established "know-how" including knowledge of "non-organic" farming "rules," as well as a general understanding of the role of nature in agriculture. In short, since practices are learned, exercised and routinized over time, moving to different practices requires moving to fundamentally different and new practice arrangements for which most farmers have no previous experience, knowledge or supportive social network (Reckwitz 2002a).

If the nature of practices is to establish "a secure and livable everyday life, where we are not compelled to do the overwhelming task of reflecting on every single act" (Gram-Hanssen 2008, 1182), their radical change, as it is often the case with converting towards organic, could be described as a temporary stage of a social practice, before re-establishing an equilibrium which arises after several years practicing organic.

How does change from non-organic to organic farming happen under those circumstances? Practices always contain the seeds of constant change (Warde 2005, 141), but something must initiate this change. From the perspective of practice theory, a specific practice-discourse initiates the adaptation and adoption of practices. In addition, the processes of change may involve bifurcations, continuous development, fragmentation, contingency, and conflict (Schatzki 2002). More specifically, the motives for converting to organic are as diverse as the barriers to converting (Lamine & Bellon 2008; Khaledi 2007; Locke 2006; Darnhofer et al. 2005) and they are linked as much with the materiality and the discourses of non-organic agriculture as with the promises and characteristics of organic practices

(Goodmann 2000). More generally, change is a discursive, communicative and material-driven break or shift in the reproduction of a practice (Reckwitz 2002b, 255). It can emerge from "... everyday crises of routines, in constellations of interpretative indeterminacy and of the inadequacy of knowledge with which the agent, carrying out the practice, is confronted in the face of the 'situation'." Obviously, social behavior often responds to stimuli and constraints from the biophysical world (Freudenburg et al. 1995, 366, in Schatzki 2010, 147 footnote 3). In other words, the material constitutes new social phenomena and vice versa. For example, excluding a formerly used materiality such as synthetic nitrogen fertilizer, creates a re-distribution of competences (cf. Shove et al. 2007, 54) and the site of the practice becomes framed by a modified order of knowledge embodied in the practices and in the new power relations which are manifested by these practices (Giddens 1984). But whenever we try to describe change of a practice, it emerges and is experienced through social practices.

The decision to convert to organic farming often follows an unforeseen path, created by the negative impact of pesticides, animal disease, decreasing soil fertility, family sickness (Jarosz 2000), a change in worldviews (ethical, religious, spiritual) or any other crisis in daily routines practices (Reckwitz 2002b, 255). Such events call into question the assumptions of current practice (c.f. Bourdieu 1993). This phase of transformation is accompanied by doubts, insecurity, disorientation, conflict and disillusion with the former ideas and practices; new questions about how to "practice" in future arise.

In the transformation period, previous practices, habits and roles in the daily farm life lose their relevance and new ones become created (see Table 3). Previous attitudes, ways of acting and thinking (Bourdieu 2001, 28) confront a new world; a new language is needed to capture new and different meanings, understandings and techniques. Farmers commonly find themselves in contradiction to their former orders of knowledge (and ways of knowing). Agricultural routines and habits, formed by family, school and education, primary socialization in agricultural communities, technical advisors, and friends, all of which shaped daily life (cf. Raphael 2004, in Jaeger and Straub 2004, 266-276), and in which the farmer was embedded, become irrelevant. A "converter" may no longer be able to learn from or adapt the practices of a "non-organic" neighbor. Moreover, the "non-organic" neighbor's practices, such as spraying pesticides or using genetically modified seeds, may now fundamentally endanger the "converter's" livelihood and well-being.

With the decision to exclude pesticides and mineral fertilizer, the farmer starts moving into a world of unknowns (Table 4). Lacking any experience in organic farming or traditional farming techniques practiced by predecessor questions arises: how to farm without mineral fertilizer, herbicides and pesticides? How to do all of this while only weakly connected to the organic movement and information, or to observations of organic practices?

In transformation, the farmer becomes the creator of a new system, developing new sets of artifacts, natural objects and social/professional/business relationships that involve new daily routines, social interactions and relationships with the farm landscape. Previous knowledge loses its significance but can also offer a starting point for adopting, developing, and establishing former and new techniques (cf. Hörning 1997, in Hepp and Winter 1997, 34). Because organic practices do not follow a recipe, the farmer often has to innovate and adapt new techniques and practices, without reference to other practices.

With limited or no experience with organic practices (Freyer 1991), the converting farmer engages in going back and forth between knowing what to do as a former non-organic farmer and embracing new and emerging approaches (Reichardt 2007, 51; Joas 1992, 239).

Type of change / practices	To exclude, to avoid	Facultative but system relevant	To adopt (modify, reduce..)..to innovate
Crop rotation	Mono-cropping	Minimum of 20% fodder legumes	New crops, crop rotation, green manure crops, legumes
Fertilization	Mineral fertilizer	Organic fertilizer management	Selected and reduced amounts of mineral fertilizers
Weed control	Herbicides	Compost management (high temperatures to eliminate seeds)	Soil tillage, mechanical weed control, crop rotation
Pest control	Majority of non-organic pesticides	Biotopes: edges and herb-grass stripes	Soil tillage, crop rotation
Varieties	Seed with pesticides, GMO	Organic seed treatment, organic varieties	new seed sources and varieties
Tillage	Deep plowing	New machines	Tillage depth modification, time periods of tillage, tillage intensity

Table 4. Quo Vadis? -Commitment of Change in plant production

Similarly, when approaching crop fertilization, the farmer must start by organizing a two-year crop rotation of nitrogen-fixing legumes. Instead of relying on a commercial product, the farmer accomplishes fertilization through the use of legumes and the creation of biomass. For the converter, this clearly represents a new practice and a fundamentally different way of thinking about and practicing fertilization. In short, crop fertilization becomes what Schatzki (2010) defines as a new practice-arrangement nexus[11]. The farmer must think and act differently about what was done as a non-organic farmer.

The material configuration of plant protection in organic as well, must be conceived in relation to soil fertility management, specific crop rotations, organic manure sprayed a certain times of the year and a specific structure of biotopes. As Schatzki (2010, 130) notes in another context, we could argue that the material configuration of organic farming takes place on the farm and in the field.

The transformation is also accompanied by tremendous changes in thinking about the farming practice. A new awareness of nature emerges, as well as new understandings of responsibilities to family, neighbors and friends. To change the system means to test, to play with new options, to leave behind routines, and to lose the stability offered by former practices (Giddens 1992, in Reichardt 2007, 59). It means being able to accept the confrontation with new heterogeneous and diverging forms of practical knowledge

[11] Practice-arrangement nexuses (linkages), according to Schatzki (2010) are social sites, which contain practices and arrangements. These practices and arrangements connect into wider nets of nexuses, in this case such as the organic community, extension services, government and financial networks, these can in turn create even wider webs of nexuses.

(Reckwitz 2003). It means learning new techniques, experimenting, contracting new relations and memberships, reading different journals, and leaning a "new language".

Because the transformation process affects all family members, the practice of converting to organic always creates family discussions. Most previous assumptions about labor, income, and decisions are "up for discussion." All these changes in orders of knowledge, power, rules and resources arise independent of the individual per se, but are phenomena of social practices that are integral to the transformation process.

The adoption of new practices also confronts the converting farmer with the need to reorganize or establish a new social network. Advisors, farmers, consumers, and institutions, the farmer has never seen before, become now of relevance (Brunori et al. 2011). With this changing and embodying of unknown materiality, things and artifacts, and social relations, inherent in practices, the farmer discovers and creates new meaning for a new farming and everyday lifestyle. By excluding the use of synthetic pesticides, herbicides and fertilizers, the farmer has to find new strategies for managing soil fertility and pests. Instead of turning to chemical products the transitioning organic farmer learns ways for managing cropping systems, soil tillage and robust varieties as part of a range of pest management strategies that must be re-defined every season. More specifically, when re-establishing biotopes around a field to foster beneficial insects, the farmer must reflect on the spatial dimensions of the field, study the living conditions of predators and to identify and establish a structural quality through which the biotope helps to fulfill the demands of various living conditions.

Non-organic and organic provide subject orders or subject cultures which are contradictory and contested (cf. Reckwitz 2008a, 80). Reckwitz underlines the strength of related cultural codes in the formation of subject forms, and the challenge to change them. Schatzki in contrast interprets stability and change dependent upon the agency of components which are configured in the arrangements, specifically that of humans (Jonas 2009, 17). Specifically in the transformation period, we argue that the capacity of individuals is more challenging than in routine situations. In part, this helps us understand the limited availability of practice-discourses on organic farming specifically at the beginning of the transformation period.

5. Concluding observations

In this chapter, we have applied practice theory as an analytic approach to help understand organic transformation as a continuing and evolving practice in which each farmer creates a new understanding and embodiment of farming materiality - soil, plants and animals - and that generates new habits and social relationships. We have also introduced the importance of understanding organic transformation as a process that is accompanied by structural changes, new orders, rules, norms and resources and new ways for the farmer to assume, restructure, but also contribute to the orders of knowledge in this type of agriculture. We have sought to identify the ways in which we might continue to draw upon practice theory to identify and generate new insights into this transformation process. Cognisant of space limitations, this paper has focused on selected aspects in the plant production sector studying an industrialized non-organic farm with a more typical organic farm. Clearly, this could be elaborated and extended to animal and poultry or vegetable production and others.

Nevertheless, our discussion does illustrate the importance of understanding, as Schatzki (2002, 174) argues, that human, nature and materiality are not separate, but together and create a material entity and contribute to a social practice. Materiality, its effects on nature, processes, practicing techniques and individuals and organizations - can be interpreted as a unit of the social and material (Schatzki 2010, 133).

We have highlighted that organic practices emerge from and rely upon natural objects, rules and ethical principles. Consequently, we find diverse farm patterns and indeterminate production processes, adapted to different and complex natural processes; and that these rules and ethical principles result from various practices. This also explains the discursive character of the social practices on an organic farm.

Principles serve partly as a "must", as norms, regulated by law and partly as recommendations. They represent the cognitive basis of the organic farming approach, which regulates the practices, and pre-forms the handling with soils and plants. The principles include the human-nature-relation, cognitively and as a matter of principle, to respect nature and to exclude synthetic chemicals from production. They represent a result of discourses based on farmers' long-term observations, practices and research. Their implementation is discursive, as the site specific conditions require their reflexive use, reinterpretation and adaptation.

The principles and the basic standards with concrete recommendations about arrangements, artifacts and practices illustrate a type of on-going practice discourse in contrast to non-organic agriculture. Neither the principles nor the basic standards are static or carved in stone. There is discourse surrounding the values and ethics (e.g. Verhoog et al. 2003), the artifacts and materialities, and the practices. There are discourses and concrete arrangements which also include evolving human-human relations, e.g., fair relation between all chain partners, embodied by a specific pricing policy or further investments by the farmer into ethical values going beyond the basic standards, e.g. social care, establishment of wildlife habitats in between the production fields or cultivation of traditional farming practices (Goessinger & Freyer 2008).

In conclusion, we find it appropriate to draw on Schatzki's (2010, 145) observation that "...one noteworthy outcome of writing histories and analyzing contemporary phenomena with these experientially resonant concepts (practice theory) is that history and the contemporary world seem less systematic or ordered and more labyrinthine and contingent than they do when described and analyzed through the conceptual armature of many other theories."

And we close with the notion that practice theory offers new insights into the complex and multi-layered process of farm transformation. It is seen as one of several relevant theoretical concepts to support the description and reflections about transformation processes in the agro-food system.

6. References

Ács, S. 2006. Bio-economic modeling of conversion from conventional to organic arable farming. PhD-thesis Wageningen University.

Adler, E. 2005. Communitarian International Relations. The epistemic foundation of International Relations. London/New York: Routledge.

Alrøe, H. F., J. Byrne & L. GloverIn 2006. CAB International 2006. Global Development of Organic Agriculture: Challenges and Prospects (eds N. Halberg, H.F. Alrøe, M.T. Knudsen and E.S. Kristensen), 76-108.

Arthur, B. 1989. Competing technologies, increasing returns, and lock-in by historical events. Economic Journal, vol. 97, pp. 642-665.

Berkes, F., J. Colding and C. Folke (Eds.) 2003. Navigating social–ecological systems: building resilience for complexity and change. Cambridge University Press, 2003.

Bertalanffy, L. von 1973. General System Theory. Foundations, Developments, Applications, fourth edition. George Braziller; New York.

Best, H. 2008. Organic agriculture and the conventionalization hypothesis: A case study from West Germany. Agriculture and Human Values (2008) 25:95–106.

Beus, C.E. & R.E. Dunlap 1990. Conventional versus Alternative Agriculture: The Paradigm Roots of the Debate. Rural Sociology 55 (4), 590-616.

Bourdieu, P.

1983. "The field of cultural production, or: The economic world reversed." Poetics, Volume 12, Issues 4-5, November 1983, Pages 311-356. http://www.sciencedirect.com/science/article/pii/0304422X83900128)(02.07.2011)

1984. Distinction: A Social Critique of The Judgement of Taste, Harvard Univ. Press.

1988. Homo Academicus. Translated by Peter Collier. Stanford University Press, 1988.

1990. The Logic of Practice. Translated by Richard Nice published in the United States by Stanford University Press, 1990.

1993. The Field of Cultural Production, Cambridge, Polity.

2001. Die Regeln Der Kunst - Genese und Struktur des literarischen Feldes, Frankfurt Am Main, Suhrkamp

Brand, K.-W. 2011. Umweltsoziologie und der praxistheoretische Zugang. in: Groß, M. (Hrsg.). Handbuch Umweltsoziologie. Wiesbaden: VS Verlag.

Brunori, G., Adanella, R. & V. Malandrin 2011. Co-producing Transition: Innovation Processes in Farms Adhering to Solidarity-based Purchase Groups (GAS) in Tuscany, Italy. Int. Journal of Soc.of Agr. & Food. Vol 18 (1) pp 28-53.

Büger, C. , & F. Gadinger 2010. Praktisch gedacht! Praxis-theoretischer Konstruktivismus in den Internationalen Beziehungen Zeitschrift für Internationale Beziehungen, 15:2. 273-302.

Callon, M. & B. Latour 1992. Don't Throw the Baby Out with the Bath School! A Reply to Collins and Yearley. In A. Pickering (Ed.) Science as Practice and Culture. Chicago, Chicago University Press: 343-368.

Cheng, Y.T. & Van De Ven, A.H. 1996. Learning the innovation journey: Order out of chaos? In: Organization Science, 7, S. 593-614

Collins, R. 2000. Ueber die mikrosozialen Grundlagen der Makrosoziologie. In Mueller, H.P. & A. Sigmund (ed.) 2000. Zeitgenoessische amerikanische Soziologie. Opladen: Leske und budrich, 99-134.Callon, M. & B. Latour 1992. Don't Throw the Baby Out with the Bath School! A Reply to Collins and Yearley. In: A. Pickering (Ed.) Science as Practice and Culture. Chicago, Chicago University Press: 343-368.

Courville, S. 2006. "Organic Standards and Certification." Pp. 201-219 in Organic Agriculture. A Global Perspective, edited by Paul Kristiansen, Acram Taji, and John Reganold. Ithaca, NY: Comstock Publishing Associates.

Cranfield, J., Henson, S., and Holliday, J. 2010. The motives, benefits and problems of conversion to organic production. Agriculture and Human Values, 27(3): 291-306

Czarniawska, B. 2007. Shadowing: And Other Techniques for Doing Fieldwork in Modern Societies. Copenhagen: Liber and Copenhagen Business School Press.

Dabbert, S. 1991. Farm planning on organic farms - theoretical issues and practical applications. Paper presented at the '6th European Association of Agricultural Economics Congress', Den Haag, September 1991.

Dahrendorf, R. 1973. Class and class conflict in industrial society. Stanford University Press, Stanford CA 1973

Darnhofer, I. 2006. Understanding family farmers' decisions. Towards a socio-economic approach, Habilitation dossier, submitted in Vienna, February 2006 to the Institute of Agricultural and Forestry Economics.

Darnhofer, I., Bellon, S., Dedieu, B., Milestad, R. 2010. Adaptiveness to enhance the sustainability of farming systems. A review. Agronomy for Sustainable Development 30(3): 545–555.

Darnhofer, I., Schneeberger, W. & Freyer, B. 2005. Converting or Not Converting to Organic Farming in Austria: Farmer Types and Their Rationale. Agriculture and Human Values, 22, 39-52.

Engel, A. 2006. Differenzierungsprozesse im Öko-Landbau: Differenzierung von Konzepten der Naturnutzung? Die Natur der Gesellschaft: Verhandlungen des 33. Kongresses der Deutschen Gesellschaft für Soziologie in Kassel 2006. Teilbd. 1 u. 2. Rehberg, Karl-Siegbert (Hrsg.) S. 1829-1840. Frankfurt am Main: Campus Verlag GmbH, 2008 [Aufsatz]

Engeström, Y. 2000. 'Activity theory as a framework for analysing and redesigning work' Ergonomics 43 (7): 960-974.

Fairweather, J. R. 1999. Understanding how farmers choose between organic and conventional production: Results from New Zealand and policy implications. Agriculture and Human Values, 16: 51-63.

Freudenburg, W. R., S. Frickel, and R. Gramling, 1995. "Beyond the Society/Nature Divide: Learning to Think about a Mountain." Sociological Forum 10 (3): 361–392.

Freyer, B. 1998. Umstellung auf IP oder Bio - eine Prognose. Agrarforschung, 5, 7, 333-336.

Freyer, B., 1991. Ökologischer Landbau: Planung und Analyse von Betriebsumstellungen. Ökologie und Landwirtschaft, Verlag Josef Margraf; Weikersheim, FRG.

Freyer, B., 1994. Ausgewählte Prozesse in der Phase der Umstellung auf den oekologischen Landbau am Beispiel von sieben Fallstudien (Selected processes during the conversion to organic agriculture, demonstrated on seven case studies), Berichte ueber Landwirtschaft, 72, 366- 390.

Freyer, B., Bingen, J., Paxton, R., Klimek, M. 2011. Practicing Soil Fertility from a Practice Theory Perspective. ISOFAR conference September 2011.

Freyer, B., Darnhofer, I., Eder, M., Lindenthal, T., Muhar, A. 2005. Total conversion to organic farming of a grassland and a cropping region in Austria-economic, environmental and sociological aspects. In: ISOFAR: 15th IFOAM Organic World Congress, 21.-23. September 2005, Australia, 308-311.

Freyer, B., Lindenthal, T., Bartel, A., Darnhofer, I., Eder, M., Hadatsch, S., Milestad, R., Muhar, A., Payer, H., Schneeberger, W.. 2002. Full conversion to organic farming in two Austrian regions, Poster In: Thompson, R. (Ed.): 14th IFOAM Conference World Congress, 21.-24. August 2002, Canada, 274.

Freyer, B., Lindenthal,T. 2002. Flächendeckende Umstellung auf biologischen Landbau - Ziele und Methoden. culterra 29/2002, 214-225; Schriftenreihe des Instituts für Landespflege der Albert Ludwig-Universität Freiburg.

Freyer, B., Rantzau, R. & Vogtmann, H. 1994. Case Studies of Farms Converting to Organic Agriculture in Germany. In: Lampkin, N. & Padel S. (eds.), The Economics of Organic Farming: An International Perspective, CAB International, Oxon, 243-263.

Giddens, A. 1984. The Constitution of Society: Outline of the Theory of Structuration. Berkeley, University of California Press.

Giddens, A. 1984. The Constitution of Society. Cambridge: Polity Press.

Giddens, A. 1992. Die Konstitution der Gesellschaft. Grundzüge einer Theorie der Strukturierung /Frankfurt/New York: Campus.

Goessinger, K. & Freyer, B. 2008. Corporate Social Responsibility and Organic Farming – Experiences in Austria. 16th IFOAM Organic World Congress, Modena, Italy, June 16-20, 2008 Archived at http://orgprints.org/view/projects/conference.html

Goodmann, D. 2000. Organic and conventional agriculture. Materializing discourse and agro-ecological managerialism. Agriculture and Human Values 17: 215-219.

Goswami ,R. & M. N. Ali 2011. Use of Participatory Exercise for Modelling the Adoption of Organic Agriculture. Journal of Extention. June 2011 // Volume 49 // Number 3

Gram-Hanssen, K. 2008. Consuming technologies - developing routines, Journal of Cleaner Production, vol. 16, 1181-1189.

Greene, C. & A. Kremen 2003. "U.S. Organic Farming in 2000-2001: Adoption of Certified Systems." Washington, DC: U.S. Department of Agriculture, Economic Research Service, Resource Economics Division.

Greene, C. & C. Dimitri 2003. "Organic Agriculture: Gaining Ground." Amber Waves 1(1): 9.

Greve, J., Schnabel, A., Schützeichel, R. 2009. Das Mikro-Makro-Modell der soziologischen Erklärung zur Ontologie, Methodologie und Metatheorie eines Forschungsprogramms, 372 S.

Guba, E.G. 1990. The paradigm dialog. Sage publications.

Guthman, J. 2004a. "Back to the Land: The Paradox of Organic Food Standards." Environment and Planning A 36(3): 511-28.

Guthman, J. 2004b. "The Trouble with 'Organic Lite' in California: A Rejoinder to the 'Conventionallisation' Debate." Sociologia Ruralis 44(3):301 - 16.

Høgh-Jensen, H. 1998. Systems Theory as a Scientific Approach towards Organic Farming. Biological Agriculture and Horticulture, 16, pp. 37-52.

Hörning, K. H. 1997. „Kultur und soziale Praxis. Wege zu einer ‚realistischen' Kulturanalyse", S. 31-46 in: Andreas Hepp und Rainer Winter (Hg.): Kultur - Medien - Macht. Cultural Studies und Medienanalyse. Opladen: Westdeutscher Verlag 1997

Hörning, K. H. 2001. Experten des Alltags. Die Wiederentdeckung des praktischen Wissens. /Weilerswist: Velbrück Wissenschaft.

IFOAM, 2009. Principles. http://www.ifoam.org/about_ifoam/principles/index.html

Jaeger-Erben, M. 2010. Zwischen Routine, Reflektion und Transformation - die Veränderung von alltäglichem Konsum durch Lebensereignisse und die Rolle von Nachhaltigkeit eine empirische Untersuchung unter Berücksichtigung praxistheoretischer Konzepte. Dissertation Humboldt Universitaet Berlin.

Jaeger, F. & Straub, J. (Eds.) 2004. Handbuch der Kulturwissenschaften. Bd. 2: Paradigmen und Disziplinen, Stuttgart, Weimar: Metzler.

Jarosz, L. 2000. Understanding agri-food networks as social relations. Agriculture and Human Values 17: 279–283

Jelsma, J. 2006. 'Designing 'moralized' products: theory and practice', in Verbeek, P.P. & A. Slob (eds), User behavior and technology development: shaping sustainable relations between consumers and technologies, Springer, The Netherlands, pp. 221-231.

Joas, H. 1992. Die Kreativität des Handelns. Frankfurt am Main: Suhrkamp.

Jonas, M. 2009. The social site approach versus the approach to discourse/practice formations. Reihe Soziologie, 92, Juli 2009. Wien: Institut für Höhere Studien. [see: http://www.ihs.ac.at/publications/soc/rs92.pdf]

Khaledi, M. 2007. Assessing the Barriers to Conversion to Organic Farming: An Institutional Analysis. http://organic.usask.ca/reports/Assessing%20the%20Barriers%20-%20Organic%20-%20Final.pdf (01.07.2011)

Lamine C. and S. Bellon, 2008. Conversion to organic farming: a multidimensional research object at the crossroads of agricultural and social sciences. A review. Agron. Sustain. Dev. 29 (2009) 97-112.

Lampkin, N. 1992. Conversion to organic farming. PhD-Thesis, University College of Wales, Aberystwyth.

Lampkin, N. 1994. Changes in Physical and Financial Performance during Conversion to Organic Farming: Case Studies of Two English Dairy Farms. In: Lampkin, N. & Padel, S. (eds.), The Economics of Organic Farming: An International Perspective, CAB International, Wallingford, 223-241.

Latacz-Lohmann, U., G. Recke & H. Wolff 2001. Die Wettbewerbsfähigkeit des ökologischen Landbaus: Eine Analyse mit dem Konzept der Pfadabhängigkeit. In: Agrarwirtschaft 50 (7): 433-438.

Latour, B. 2005. Reassembling the social. Oxford: OUP.

Lockeretz W. 1995. Organic farming in Massachusetts: An Alternative Approach to Agriculture in an Urbanized State, Journal of Soil and Water Conservation, Vol. 50 (6), 663-667, PadelADEL S. (2001). Conversion to Organic Farming: A Typical Example of the Diffusion of an Innovation? Sociologia Ruralis, Vol. 41 (1), 40-61.

Lorand, A. C. 1996. Biodynamic Agriculture—A paradigmatic Analysis. The Pennsylvania State University, Department of Agricultural and Extension Education.PhD Dissertation.

MacRae, R. J., S. B. Hill, J. Henning & G. R. Mehuys 1989. Agricultural science and sustainable agriculture: a review of the existing scientific barriers to sustainable food production and potential solutions. 6 173-219.

Meadows, D. 2008. Thinking in systems - a primer. Sustainability Institute (Earthscan)

Michelsen, J. 2001. 'Recent development and political acceptance of organic farming in Europe', Sociologia Ruralis, Vol. 41, No. 1, pp.3–20.

Nicolini, D. 2009. Zooming in and out: studying practices by switching theoretical lenses and trailing connections. Organization Studies, 30(12).

Noe, E. & & Alroe, H.F. 2006. Combining Luhmann and Actor Network Theory to See Farm Enterprises as Self-organizing Systems. Cybernetics And Human Knowing. Vol. 13, no. 1, pp. 34-48

Noe, E. & Alroe, H.F. 2003. Farm enterprises as self-organizing systems: a new transdisciplinary framework for studying farm enterprises? International Journal of Sociology of Agriculture and Food 11(1): 3–14 (2003)

Odening, M., O. Mußhoff & V. Utesch 2004. Der Wechsel vom konventionellen zum ökologischen Landbau: Eine investitionstheoretische Betrachtung Adoption of organic farming - the impact of uncertainty and sunk costs. Agrarwirtschaft 53 (2004), Heft 6

Ortner, S. 1984. 'Theory in anthropology since the 60s' Comparative Studies in Society and History 26(1): 126-166.

Padel, S. 2001. Conversion to organic farming: A typical example of the diffusion of an innovation? Sociologia Ruralis, Vol 41, No 1.

Pali, P. N., S.K. Kaaria, R.J. Delve, B. Freyer, 2011. Can Markets Deliver the Dual Objectives of Income Generation and Sustainability of Natural Resources in Uganda? Short title: Sustainable Markets with Soil Management. African Journal of Agricultural Research Vol. 6(10)

Parsons, T. 1970. The Social System. London: Routledge & Kegan Paul Ltd.

Preda, A. 1999. The Turn to Things: Arguments for a Sociological Theory of Things. The Sociological Quarterly, Vol 40, Nr 2, 347-366.

Pouliot, V. 2008. The Logic of Practicality: A Theory of Practice of Security Communities, in: International Organization 62: 2, 257-288.

Rammert, W. 2008. Where the action is: Distributed agency between humans, machines, and programs" In: Seifert, U, J.H. Kim & A. Moore (eds.), Paradoxes of interactivity: perspectives for media theory, human-computer interaction, and artistic investigations, 2008. Transcript verlag.

Rantzau, R., Freyer, B. & Vogtmann, H. 1990. Umstellung auf oekologischen Landbau (Conversion to Organic Agriculture), Reihe A: Angewandte Wissenschaften, Heft 389, Landwirtschaftsverlag GmbH, Muenster-Hiltrup.

Rasche, A. & R. Chia 2009. Researching Strategy Practices: A Genealogical Social Theory Perspective. Organization Studies 2009; 30; 713-735.

Reckwitz, A. 2008a. Subjekt/Identität: Die Produktion und Subversion des Individuums. In: Andreas Reckwitz / Stefan Möbius (Hg.): Poststrukturalistische Sozialwissenschaften, Frankfurt am Main: Suhrkamp 2008.

Reckwitz, A. 2008c. Die Kontingenzperspektive der >Kultur<. Kulturbegriffe, Kulturtheorien und das kulturwissenschaftliche Forschungsprogramm. pp. 15-45 In: Reckwitz, A. 2008c. *Unscharfe Grenzen - Perspektiven der Kultursoziologie.* Bielefeld: transcript

Reckwitz, A. 2006. Das hybride Subjekt. Weilerswist: Velbrück.

Reckwitz, A. 2002a. The status of the 'material' in theories of culture. From 'social structure' to 'artefacts'', Journal for the Theory of Social Behaviour, vol. 32, no. 2, pp. 195-217.

Reckwitz, A. 2002b. Toward a theory of social practices: a development in culturalist theorizing', Journal of Social Theory, vol. 5, no. 2, pp. 243-63.

Reckwitz, A. 2003. Grundelemente einer Theorie sozialer Praktiken. Eine sozialtheoretische Perspektive. Zeitschrift für Soziologie, 32 (4), 282- 301.

Reckwitz, A. 2005. Kulturelle Differenzen aus praxeologischer Perspektive. Kulturelle Globalisierung jenseits von Modernisierungstheorie und Kulturessentialismus. In Ilja Srubar, Joachim Renn & Ulrich Wenzel (Eds.), Kulturen vergleichen. Sozial- und kulturwissenschaftliche Grundlagen und Kontroverse (pp.92-111). Wiesbaden: VS.

Reckwitz, A. 2008b. Praktiken und Diskurse. In: Kalthoff, H.erbert, S.tefan Hirschauer und G.esa Lindemann (Hrsg.), Theoretische Empirie - Zur Relevanz qualitativer Forschung. Frankfurt/M.: Suhrkamp, 188-209.

Reichardt, S. 2007. Praxeologische Geschichtswissenschaft. Eine Diskussionsanregung. Sozial. Geschichte, 22, 3, 43-65.

Reissig, R. 2009. Gesellschafts-Transformation Im 21. Jahrhundert: Ein Neues Konzept Sozialen Wandels, Vs Verlag.

Rogers, E. M. 1995. Diffusion of innovations (4th ed.). New York: Free Press.

Ryan B. & N. C. Gross 1943, "The Diffusion of Hybrid Seed Corn in Two Iowa Communities," Rural Sociology 8 (March): 15.

Schatzki, T. R. 2002. The Site of the Social: a Philosophical Account of the Constitution of Social Life and Change, The Pennsylvania State University Press, Pennsylvania [USA].

Schatzki, T. R. 2010. Materiality and Social Life, Nature and Culture 5 (2): 123- 149

Schatzki, T. R. 1997. Practices and Actions; A Wittgensteinian Critique of Bourdieu and Giddens. Philosophy of The Social Sciences, 27, no. 3, pp. 283-308.

Schatzki, T. R. 1996. Social practices: A Wittgensteinian approach to human activity and the social, Cambridge, MA: Cambridge University Press.

Schatzki, T. R. 2005. 'The Sites of Organizations'. Organization Studies 26: 465-484.

Schneeberger, W., Darnhofer, I. & Eder, M. 2002. Barriers to the Adoption of Organic Farming by Cash-Crop Producers in Austria. American Journal of Alternative Agriculture, 17, 24-31.

Shove, E. & Pantzar, M. 2005. Consumers, Producers and Practices - Understanding the invention and reinvention of Nordic walking. Journal of Consumer Culture, 5 (1). 43-64.

Shove, E. & Pantzar, M. 2007. Recruitment and reproduction: the careers and carriers of digital photography and floorball. Journal of Human Affairs, 17, 154-167.

Shove, E., M. Watson, M. Hand & J. Ingram, 2007. The Design of Everyday Life, Oxford/New York: Berg.

Strengers, Y. 2010. Conceptualising everyday practices: composition, reproduction and change. Working Paper No. 6, Carbon Neutral Communities Centre for Design, RMIT University. 1-21.

Thevenot, L. 2001. Pragmatic regimes governing the engagement with the world. In: Schatzki, T. R., Knorr-Cetina, K. & E. von Savigny (ed.) 2001. The Practice Turn in Contemporary Theory. London: Routledge, 1-14.

Tranter, R., Bennett, R., Costa, L., Cowan, C., Holt, G., Jones, P., Miele, M., Sottomayor, M. & Vestergaard, J. 2009. Consumers' Willingness-To-Pay For Organic Conversion-Grade Food: Evidence From Five EU Countries. Food Policy, 34, 287-294.

Tress, B. 2001. Converting to organic agriculture - Danish farmers' views and motivations. Geografisk Tidsskrift, Danish Journal of Geography 101: 131-144.

Turner, J. 1991, Social Influence, Open University Press, Buckingham, UK.

United States Department of Agriculture. 2002. "National Organic Program Final Rule."

Verhoog H., Matze M., Lammerts van Bueren E. & Baars T. 2003. The role of the concept of natural (naturalness) in organic farming. Journal of Agricultural and Environmental Ethics, 16, 29- 49.

Warde, A. 2005. Consumption and Theories of Practice. Journal of Consumer Culture. Vol 5(2): 131–153.

Wehinger, T., Freyer, B., Hoffmann, V. 2002. Stakeholder Analysis in the conversion to organic farming. Pre-proceedings to the Fifth IFSA European Symposium on Farming and Systems Research and Extension - Local Identities and Globalisation, Florence, Italy (8.-11.04.2002) P 767-776. In: Università degli Studi di Firenze, CeSAI, Istituto Agronomico per l'Oltremare (Eds.): Fifth IFSA European Symposium on Farming and Systems Research and Extension - Local Identities and Globalisation.

Westermayer, T. 2007. Soziale Praktiken, un-published manuscript, Freiburg.

10

Sustainable Food System – Targeting Production Methods, Distribution or Food Basket Content?

Markus Larsson[1], Artur Granstedt[2] and Olof Thomsson[3]
[1]*Stockholm University and Mälardalen University,*
[2]*Södertörn University,*
[3]*The Biodynamic Research Institute,*
Sweden

1. Introduction

Agriculture is the single most important contributor to the eutrophication of the Baltic Sea. It is responsible for 59% of the anthropogenic nitrogen and 56% of the phosphorous emissions (HELCOM, 2005). A second important source of nutrient emissions is at the other end of the food system - emissions from municipal waste-water treatment plants and from private households. Addressing different aspects of the food system is thus crucial for the Baltic Sea environment. To tackle eutrophication both nitrogen and phosphorous loads should be reduced (MVB, 2005). This can be achieved if emissions from the food system are reduced, e.g. by closing the nutrient cycle from soil to crop and back to agricultural soil (Diaz and Rosenberg, 2008). Granstedt (2000) finds that the high surplus and emissions of nitrate and phosphorous in Swedish agriculture is a consequence of specialized agriculture with its separation of crop and animal production. Similar findings are reported from different parts of Europe (Brower et al., 1995). About 80% of cropland in Sweden is used for fodder production but the animal production is concentrated to a limited number of specialized animal farms. Manure, with its contents of nutrients from the whole agriculture area, is today concentrated on only 20% of the Swedish arable land (Statistics Sweden, 2011). This results in high nutrient surplus and load of nitrogen and phosphorus from these areas. Granstedt (2000) concludes that the emissions can be limited by combining best available agricultural technology with increased recycling of nutrients within the agricultural system trough integration of crop and animal production - ecological recycling agriculture (ERA). This facilitates an efficient use of the plant nutrients in farm yard manure. Other studies of nutrient balances comparing farming systems and lifecycle assessment report similar observations (Halberg, 1999; Myrbeck, 1999; Steinshamn et al., 2004; Uusitalo, 2007). The potential of reduced nutrient emissions trough ERA was confirmed in case studies on local organic farms around the Baltic Sea (Granstedt et al., 2008; Larsson and Granstedt, 2010).

Carlsson-Kanyama (1999) found that greenhouse gas emissions could be reduced by local and organic food production due to shorter transportation. Similar results are reported in a compilation of studies (FiBL, 2006) and in studies of local production and processing in Järna, Sweden (Wallgren, 2008). According to Carlsson-Kanyama et al. (2004) the reductions are not significant unless local distribution becomes more efficient.

1.1 Aim and research questions

The aim of this study was to investigate how much the environmental impacts could be reduced by various changes in the governance of food systems. The main questions investigated were; the importance of food production methods, the importance of transport and processing systems of food, and the impact of different food consumption profiles. By examining this we can also answer what effort would give the most environmental benefits to society. The environmental impacts assessed were potential emissions of nitrogen (risk for eutrophication), global warming impact and use of primary energy in the agricultural production, transporting and processing parts of the food system. To define an alternative food basket and to calculate costs borne by households a consumer survey was carried out. The aim with the survey was to provide information on what a food basket of environmentally concerned residents consists of and costs, in one case study site.

The environmental impact of different food choices has gained increased attention in policy documents. In its "Strategy for sea and coast free from eutrophication" the Swedish Environmental Advisory Council states that major reductions in nitrogen emissions from agriculture are possible with changed consumption profiles (MVB, 2005). Similar recommendations are found in a Government Commission Report on sustainable consumption where increased shares of vegetables as well as local and organic food as means to achieve "sustainable consumption" are discussed (SOU, 2005).

Governance of ecosystems or natural resources is often more efficient if several sectors (horizontal collaboration) and several levels (vertical collaboration) are involved in the process (Low et al., 2003). When addressing the food system different sectors include the production and the consumption sides and levels of decision range from farmers and consumers to municipal and governmental agencies and the EU. Local stakeholder collaboration ensures that several objectives (ecological, social, and economic) are addressed. Such horizontal collaboration involves public agencies as well as NGO:s. Vertical collaboration, or multilevel social networks (Adger et al., 2005) on the other hand, is crucial for enhancing social and ecological resilience (Folke et al., 2005; Dietz et al., 2003). Collaboration and different aspects of institutions in the food sector have been studied by e.g. Carlsson-Kanyama et al. (2004) and Larsson et al. (2007). This paper draws on results obtained in the two EU financed projects BERAS and GEMCONBIO[1].

2. Methodology

Environmental effects of different farming, processing and distribution regimes and different consumption profiles were studied, using primary and secondary data compiled from own studies, literature and official statistics. First we present an overview of the methodology used. More details are given in the following sub-chapters.

Two farming systems, Swedish average 2000-2002 and the system of Ecological Recycling Agriculture (ERA, see Box 1 below) farms 2002-2004 were used for comparison of their respective environmental impacts. Two food consumption profiles (food baskets) were used for comparison of the importance of our choices of food. Data on consumption patterns from a national consumer survey (Swedish Board of Agriculture, 2004) represent the Swedish average food basket. Data obtained in the consumer survey carried out represent an

[1] Baltic Ecological Recycling Ecology and Society, www.jdb.se/beras and Governance and Ecosystem Management for the Conservation of Biodiversity, www.gemconbio.eu.

alternative "eco-local" food basket. Two processing and transportation scenarios were used. "Conventional", mainly large-scale food processing with long-distance transports (data in most cases earlier reported in Carlsson-Kanyama et al. (2004)) and "Local" based on data collected from businesses in Järna, a rural community south of Stockholm.

Four scenarios representing different combinations of agricultural production systems, food consumption profiles, and food processing and transportation systems were combined to answer the research questions. The scenarios are:

1. Conventional scenario - average Swedish food consumption profile, average Swedish agriculture, and conventional food processing and transports.
2. Conventional consumption from ERA farms - average Swedish food consumption profile, ERA farms, and conventional food processing and transports.
3. Local consumption from ERA farms - average Swedish food consumption profile, ERA farms, and local (small-scale) food processing and transports.
4. More vegetarian and local consumption from ERA farms - an alternative food consumption profile (e.g. less and different kinds of meat), ERA farms, and local (small-scale) food processing and transports.

To achieve a match between consumption and production in the second and third scenarios when ERA farms produce the food for the average Swedish food consumption profile it was necessary with the assumption that the consumption volumes of ruminant meat (beef and lamb) and monogastric meat (pork and poultry) can be exchanged depending on the higher share of ruminant meat production on the documented ERA farms.

In all 12 organic, or ERA, farms were studied. The farms were selected to be representative for the main farming conditions and production types in Sweden. They were studied during the years 2002-2004. The studied farms are spread over central and southern Sweden with a concentration (6 out of 12) in the Järna region. Farm characteristics, production data and use of resources were inventoried using interviews and farm accounts. Corresponding data for average Swedish agriculture was obtained from Statistics Sweden (2005).

Ecological Recycling Agriculture (ERA) is a local organic agriculture system based on local and renewable resources. ERA produces food and other agriculture products according to the following basic ecological principles (Granstedt, 2005):

1. Protection of biodiversity.
2. Use of renewable energy.
3. Recycling of plant nutrients.

In consequence with these principles an ERA farm is defined as an organic (ecological) managed farm according the IFOAM standards[2] with no use of neither pesticides nor artificial fertilizers (IFOAM principles 1 and 2) and with the additional condition of a high rate of recycling of nutrients based on organic, integrated crop and animal production. A higher degree of internal recycling within the system enables reduced external input of nitrogen. Nitrogen requirements are covered through biological nitrogen fixation of mainly clover/grass leys. There is only a limited deficit of phosphorus and potassium in the input and output balance according to previous studies

[2] IFOAM, International Federation of Organic Agriculture Movements. The standards are described at www.ifoam.org.

(Granstedt, 2000). The greater part of the minerals is recycled within the farm in the manure. The limited net export of phosphorus and other nutrients seems to be compensated by the weathering processes in most soils and a recycling of food residues could further decrease these losses from the system (Granstedt, 2000; Granstedt and Kjellenberg, 2011). The strive to be self-sufficient in fodder limits the number of animals per hectare. In reality, however, some smaller amounts of imported inputs (seeds, fodder and rock powder for soil improvements) can be necessary depending on variation in yield level between different years. An external fodder rate of a maximum of 15% of total fodder and an animal density of <0.75 animal units/ha were used as criteria for selecting ERA-farms (Granstedt, 2005). An animal unit (au) is defined as one dairy cow, or two young cows, or three sows, or ten fattening pigs, or 100 hens. By following these principles nutrient in manure does not exceed what can be utilised by crops during the crop rotation in the same system. Each single farm does not need to function as a closed system. Farms in the same region with complementing production could cooperate and together function as a recycling farming system in terms of fodder and manure, but regional specialisation of production is problematic. The studies are based on calculated surplus and emissions of reactive nitrogen and surplus of phosphorus compounds from the agriculture–society system according to methods developed by Granstedt (1995; 2000; 2005).

[i]Adapted from Larsson and Granstedt (2010).

Box 1. Principles of ecological recycling agriculture systems[i]

2.1 The consumer survey and calculating an alternative food basket

A consumer case study was carried out in Järna, a community of around 7500 inhabitants. Järna is part of Södertälje municipality, located in Stockholm County. Järna was chosen because there are numerous biodynamic and organic farms and market gardens in the area that serve the local market and a well developed consumer network linked to these farms, i.e. it was possible to find a group of environmentally concerned consumers that was willing to take part in the survey. There are also several food processing industries like a mill and bakery (with both a local and national market), a farm-size dairy and a farmer cooperative selling organic vegetables and meat. For a more detailed presentation of the site see Haden and Helmfrid (2004) and Wallgren (2008).

The families participating in the survey recorded their food purchases for two two-week periods in 2004; one in winter/spring (when local products are scarce) and one in late summer/early autumn (when local products are easy available). The periods were chosen in order to get representative results for the yearly consumption. Information on the amount, price, origin and environmental brand (e.g. Demeter or KRAV)[3] of all food products was recorded either on the detailed receipts or on specified lists supplied. Since the matter of concern in the study was environmental impacts from food and local production, and the studied farms only produce "real food", only these types of products were included in the consumer study. Products such as sugar, candy and beverages were thus not included.

After the recording period, the families were interviewed about their food choices, food consumption and food purchasing habits. The amounts of different products purchased

[3] KRAV is the certifying organisation of organic products in Sweden and Demeter is the equivalent for biodynamic products. See www.krav.se and www.demeter.nu.

during the measured four weeks were then extrapolated to get values for consumption during the whole year. For some comparisons to Swedish average figures, the results for the Järna consumers were also extrapolated to cover meals eaten outside the home, on average 16% of their meals. Apart for content of an alternative food basket the consumer survey also estimated the household cost for this. In all 49 individuals in 15 households took part in the survey. The families were invited to take part in the survey through local food and environment organisations. No formal socio-economic stratification of the families were performed but the general picture obtained during the interviews was that the families well represent the Swedish society. Considering the low number of participants the results obtained is to be considered a special case (scenario 4) and wider implications should be interpreted carefully.

2.2 Calculating agriculture area and environmental impacts of food baskets

The annual environmental impacts of a food basket from ERA farms is calculated from data on consumption (kg per capita and year) of different food product categories, and data on the annual agricultural production (kg per ha). In this study only agriculture land in Sweden was included in the calculations of environmental impacts in terms of nutrient surplus, use of energy and emissions of green house gases. The external fodder including nutrients in imported fodder was however included in the input resources in the nutrient balances.

Products in the food basket were calculated back to the original (primary) amounts of agricultural products produced for human consumption. These included weight of crops harvested for food consumption (kg), living weights for animals going to slaughter (kg) and delivered milk (kg). Calculations for the different products from the average Swedish agriculture and from the ERA agricultures represented by the 12 ERA prototype farms were done using the equation:

$$O = C*cf \tag{1}$$

where C is the amount of a food stuff (kg), cf is a conversion factor for converting a foodstuff back to the weight of the original agricultural product (O). The conversion factors for the different foodstuffs were based on the database FAOSTAT (2004) and complemented with information from Saltå Mill (bread and cereal products) (Gustavsson, 2003) and Svensk Mjölk (dairy products) (Pettersson, 2005). Since the production levels and environmental impacts differed greatly between the farms and farming systems, the original food production in kg was converted to area (ha) in order to get proper results.

The products from the ERA farms included in the food baskets were grouped into seven agricultural categories: Potato products (O_1); Grain products (O_2); Root crops (O_3); Vegetable products (O_4); Milk products (O_5); Meat from ruminant animals (O_6); Meat from monogastric animals (O_7). Characteristic for the Swedish ERA-farms is that they all integrate crop and animal production. However, it was possible to group the farms according to production following four groups with the dominant product named first (bold):

1. **Potatoes,** root crops, vegetable products, bread grain and milk (2 farms)
2. **Milk,** meat and bread grain (6 farms)
3. **Pork,** poultry, egg and cereals (2 farms)
4. **Ruminant meat** and cereals (2 farms)

The farms in group 1 were more diverse and produced a broad spectrum of agricultural products. The farms in group 4 were more specialised and more extensive. To calculate the agriculture area needed and the environmental impacts of food basket scenario 2 and 3 two methods was used:

1. The production (kg per ha) of the products from the four farm groups was combined so they together cover the annual demand of the seven consumption categories in the annual food baskets. The environmental impacts of the food basket were calculated from the average impact of the four farm categories respective.
2. The average production (kg per ha) of products from all the 12 ERA farms was combined so they together cover the annual demand of the seven consumption categories in the annual food baskets. The environmental impacts of the food basket were calculated from the average impact of the all 12 studied ERA farms.

2.2.1 Nutrient surplus

The method for calculating nutrient balances follows those described in Granstedt (2000) and Larsson and Granstedt (2010). The potential emissions of nitrogen were defined as the difference between total input of nitrogen to the farm and the export from the farm in form of agricultural products (meat, milk, grain and horticultural products) (Granstedt et al., 2004). A steady state of the total nitrogen content is assumed. An increased content of Soil Organic Matter (SOM) has however been observed in several studies of organic farms (Granstedt and Kjellenberg, 2008; Hepperly et al., 2006; Mäder et al., 2002) which implies that real losses of nitrogen can be lower than the observed surplus in the nutrient balances.
The potential nitrogen emissions from each farm group as a part of the total load from one food basket was calculated using the equation:

$$A_{i\ N\text{-surplus}} = A_i * A_{i\ N\text{-surplus/ha}} \tag{2}$$

where $A_{i\ N\text{-surplus}}$ is the N-surplus (kg) from the area used for one food basket from farm group i, i=1-4, A_i is the area for farm group i and $A_{i\ N\text{-surplus/ha}}$ is the average N-surplus per ha from the ERA farms included in farm group i.
The nitrogen surplus of one food basket was calculated using the equation:

$$A_{N-\text{surplus}} = \sum_{i=1-4} A_{i\ N-\text{surplus}} + A_{\text{diff}\ N-\text{surplus}} \tag{3}$$

where $A_{N\text{-surplus}}$ is the total N-surplus from the area used for food production per capita (i.e. food basket), and $A_{\text{diff}\ N\text{-surplus}}$ is the summarised residual value of N-surplus for the seven food product categories converted to area (ha). Both primary and official data were used in the calculations. The same procedure was also used for global warming potential and consumption of primary energy resources.

2.2.2 Global warning impact and energy use

The assessment of global warming impact and primary energy use followed the principles of life cycle assessment (LCA) methodology (Lindfors et al., 1995), although a complete LCA was not made due to the complexity of the systems studied. The LCA methodology is primarily designed for assessment of single products, but the structure of the methodology can also be used for larger systems. Here assessments where first made separately for the agriculture, the processing and the transportation systems. There after these results where used in assessment of the scenarios. Compared to a complete LCA the steps being omitted include; the assessment of several impact categories, some minor system parts of the data inventory, and a full description of system borders.

Following the LCA process, a life cycle inventory (LCI) inventorying data concerning direct and indirect energy use and resource consumption were performed in all vital parts of the system under study. For "conventional" agriculture, processing and transportation secondary data were used. For the "alternatives" the 12 ERA farm and food processing and transporting business in the Järna area were used. The data were then grouped into impact categories, where one emission may contribute to several categories. This study assess the impact categories "Global warming impact" and "Use of resources - fossil energy", since these two impacts are closely linked to each other and because they were judged to be the most important ones.

Global warming impact was assessed using global warming potentials (GWP), where all impacting emissions are transformed into CO_2-equivalents. Only direct impacting gases were inventoried, i.e. CO_2, CH_4 and N_2O. The GWP of CH_4 and N_2O correspond to 23 and 296 CO_2-equivalents respectively. Of the different time-spans suggested by IPCC (2001) the 100-year perspective was chosen. The inventory of energy use included two categories of energy carriers - electricity and fossil fuels. These were re-calculated as primary energy, i.e. the energy used was converted to primary energy resource equivalents, measuring the consumption of energy resources in the lifecycle of the energy carriers. Transmission losses in the distribution net (7%), pre-combustion energy consumption for fuels and efficiency in e.g. hydropower and nuclear power are included in the assessment (Lundgren, 1992). This made it possible to compare scenarios and activities using mainly electricity with those using mainly fossil fuels. The results are based on data from the 12 studied Swedish ERA farms. Whether these perform better or worse than other organic farms is not investigated.

3. Results

Environmental impacts of conventional Swedish food production of an average food basket (Scenario 1) is compared with food produced with ERA-methods (Scenario 2); food produced with ERA-methods and processed locally (Scenario 3), and finally with an alternative food basket with less meat and more vegetables produced with ERA-methods and processed locally (Scenario 4). First, the results of the household survey are presented.

3.1 The household survey

When studying the results from the Järna survey there are some evident differences between the consumption patterns of the investigated households and the Swedish average. An average of 73% of the weight for what is considered 'real food' (sugar, candy, beverages etc. not included) was reported as being organic, or ecological, in the alternative food consumption profile, the "eco-local" food basket. In comparison with the national average of 2.2% the figure is very high. Some of the Järna consumers mentioned that they would have bought more eco-food if it was available and not too expensive. The portion of locally produced food purchased by the investigated households was found to be substantial for some product groups, e.g. 56% for cereals and 49% for beef and lamb. On average 33% was reported being local and organic. It is not possible to compare with national averages concerning local food but it is reasonable to assume that the average share is very low.

Other important characteristic for the eco-local food basket were the substantially lower shares of meat and potatoes (75% respectively 57% less) and the higher vegetable consumption (100% more), see Table 1. When looking at more detailed product groups some interesting differences become apparent. Although there is no difference in cereal products as a group it can easily be seen that these households seem to bake more of their bread at home (buy more flour but less bread). They also eat more groats and flakes, which is in accordance with the higher consumption of yoghurt and other fermented dairy products and prefer butter to the more processed margarine.

Product group	Sweden average		Järna survey 2004[i]				
	total[ii]	eco[iii]	total	eco		eco-local[iv]	
	kg	%[v]	kg	kg	%	kg	%
Cereal products	103	1.6	103	81	78	58	56
Potatoes	54	3.3	23	22	96	9	38
Root crops	9	9.9	42	39	92	17	40
Vegetables, veg. products and legumes	58	2.0	98	64	66	29	30
Milk products	168	5.1	199	162	81	72	36
Meat ruminants (beef and lamb)	12		7	5	70	4	49
Meat monogastrics[vii] (pork and poultry)	28	0.8[vi]	2	1	48	1	28
Other meat and mixed meat products	37		5	3	62	2	41
Egg	9	9.7	7	6	88	2	22
Fish and fish products	18	--[viii]	5	0	3	0	0
Fat	13	2.7	15	6	42	0	0
Fruit, berries, nuts and seeds	63	2.6	80	39	48	2	3
Total 'real food', excl. sugar, candy, beverages etc.	572	2.2	584	428	73	194	33

[i] compensated for meals eaten outside home
[ii] Swedish average 2002 (Swedish Board of Agriculture, 2004)
[iii] certified KRAV, and/or Demeter
[iv] produced in Järna district and certified according to KRAV and/or Demeter
[v] % of expenditures per product group
[vi] % of all meat and meat products
[vii] In scenarios 2 and 3, the consumption of ruminant and monogastric meat was swapped in order to fulfill crop rotation demands and a minimum of 40% clover/grass leys in agriculture. Ruminants, beef cattle and sheep, are the only animals that can digest crops like grass and clover. Monogastric animals like pigs and poultry are mainly fed with grain.
[viii] not possible to certify at that time

Table 1. The share of ecological and local food purchases, kg per capita and year, and % of weight.

3.1.1 Household food expenditures

In Järna the investigated households spend more money on food than the average Swedish household. The mean value for food expenditures per household was 5833 €/household/year in the monitored households, while the Swedish average household expenditures was 3376 €, alcoholic beverages and restaurant meals not counted (Statistics

Sweden, 2004). However, when calculated per consumption unit[4] (CU) the difference is smaller, 2600 €/CU/year in Järna compared to 2100 € for the Swedish average CU, 24% higher expenditures. Whether this is a result of these families really giving higher priority to food or a result of the socio-economic status of the studied households was not investigated.

	€/CU	€/person/year	€/household/year
Järna	2584	1800	5833
Swedish average	2084	1600	3376

Table 2. Expenditures on food.

The method used in this consumer survey has some potential limitations. Purchasing patterns may be distorted and no information on the distribution of foods within households is normally obtained (Cameron and van Staveron, 1988). One problem is the possible lack of information about whether a product is never purchased or whether it simply was not purchased during the recorded weeks (Irish, 1982). Bulk purchases make it more difficult to estimate annual food expenditures than if the consumers acquire all or part of their food in relatively small quantities once or several times per week (Pena and Ruiz-Castillo, 1998). However, when the families were interviewed and their purchase diaries and collected receipts checked, information on the above issues was received.

3.2 Nutrient surplus and land use

Table 3 presents base data and the calculated nitrogen surplus in agriculture based on the four production-type groups of ERA farms (potatoes and root crop; milk and meat; pork, poultry and cereal; ruminant and cereal) compared to the average Swedish agriculture, Scenario 1.

The results from scenarios 2 and 3 (conventional consumption from ERA farms and locally produced consumption from ERA farms) are the same because different processing and transport systems have no influence on nutrient surplus in agriculture. The surplus of nitrogen (total and per capita) in the scenario based on ERA-farms, with the same total meat consumption (but with a higher share of ruminant meat), is reduced with 37% compared to the same food being produced by the average Swedish agriculture with the calculation based on the four categories of farms (calculation according method 1 described in 2.2) and 18% compared the average Swedish agriculture with the calculation based on the average surplus on all the 12 Swedish ERA farms (calculation according method 2 described in 2.2). The nitrogen surplus per hectare is also very low in ERA production. This calculation is based on the total farm gate balance including emissions of ammoniac from the animal production[5]. However, scenarios 2 and 3 require having 4.76 million hectares under agriculture production, compared to the 2.45 million hectares arable land of today. This larger

[4] CU = Consumption Unit, a measure that compensates for household structure and the ages of the household members to allow for more relevant comparisons of consumption between different household types.

[5] Calculating the nitrogen surplus as field balances would result in greater differences. Field balances, i.e. excluding the emissions from animal production, give 70 – 75 % lower surplus of nitrogen from soil and corresponding losses to the water system compared to the average Swedish agriculture (Granstedt et al., 2008).

area was partly a result of a lower production on organic farms and mainly a result of a higher share of ruminant meat (70% compared to 30% in conventional production) which requires more arable land compared to when producing pork and poultry. In Table 3 both per hectare and per capita figures are presented. The latter figures are the more important ones.
Figure 1 shows the results for nitrogen surplus in diagram form for the sake of comparison to the results presented in the following section (calculated according to method 1).

	Scenario 1. Average Swedish cons. & agri. 2000-02[i]		Scenario 2 and 3. Swedish consumption & ERA farms 2002-04		Scenario 4. Eco-local consumption & ERA farms 2002-04	
		%		%		%
Agriculture area, million ha in Sweden	2.45	100	4.76	194	1.70	69
Agriculture area, ha/capita	0.27	100	0.53	194	0.19	69
N-surplus, kg/capita	22	100	14	63	8	36
(Method 2)	(22)	(100)	(18)	(82)	(10)	(45)
N-surplus, kg/ha	80	100	26	32	42	52
N-surplus, million kg in Sweden	196	100	123	63	71	36

i Adapted from Statistics Sweden (2005). Only arable land in production is counted.

Table 3. Agricultural area required and nitrogen surplus for three scenarios: Swedish average (mainly conventional) agriculture, ERA farms producing the average Swedish food-basket, and ERA farms producing an alternative (ecological and more vegetarian) food-basket. In Scenario 2, 3 and 4 all agricultural production is turned into ERA. Figures within brackets represent are calculated with method 2, see section 2.2. Other results are obtained using method 1.

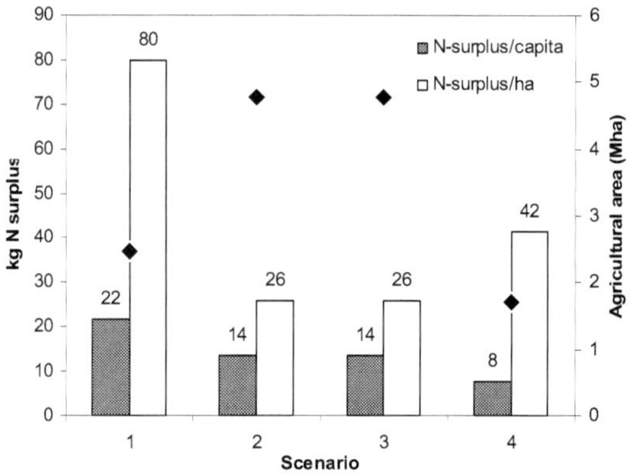

The black diamonds represent the required area for agricultural production, *million hectares.* 1) Conventional; 2) Conventional consumption from ERA farms; 3) Local consumption from ERA farms; 4) More vegetarian consumption from ERA farms.

Fig. 1. N-surplus in four scenarios, *kg N per capita and kg N per ha.*

Scenario 4 assumes more vegetarian food consumption produced on ERA-farms. In this scenario, the area of agricultural arable land would *decrease* by slightly more than 30% to 1.7 million hectares. And most important, the nitrogen surplus would decrease by 64% or 55% of today's level, depending on if method 1 or 2 is used for calculating the surplus.

3.3 Global warming impact and primary energy resources consumption

Figure 2 and Figure 3 present the results for global warming impact (measured as GWP in CO_2-equivalents) and consumption of primary energy resources (measured in MJ primary energy resources). Here, four scenarios are included as the different systems of processing and transportation are also compared. The trends are similar to those for nitrogen surplus in both cases. However the differences between the scenarios are smaller for the GWP. Changing to ERA-production (Scenario 2) resulted in a 10% reduction in GWP, from 1000 to 900 kg CO_2-equivalents with the calculation based on the four categories of farms (calculation according method 1 described in 2.2). The very low per-hectare results in Scenario 2 and 3 are a result of these scenarios requiring a very large (and unrealistic) area under agriculture production. In Scenario 4 (ERA-production, local processing and distribution and a more vegetarian food profile) the GWP is reduced with 40% compared to Scenario 1.

For the primary energy resources consumption the relation is almost exactly the same as for nitrogen surplus. The use of primary energy for the food consumption is reduced with 44% per capita with the calculation based on the four categories of farms (method 1) with food from ERA agriculture with a traditional diet but with a large part of the monogastric meat replaced by ruminant meat (Scenario 2). Reduced meat consumption with 75%, would reduce the primary energy use with an additional 40%, or in total 67% (Scenario 4).

Processing food locally (and the resulting shorter transports) has some impact on the GWP but almost no impact on the primary energy resources consumption (Scenario 2 vs. Scenario 3). The latter can partly be explained by the choice of energy carriers (fossil fuels vs. electricity) in the food processing industries and by very inefficient meat transports in the studied case.

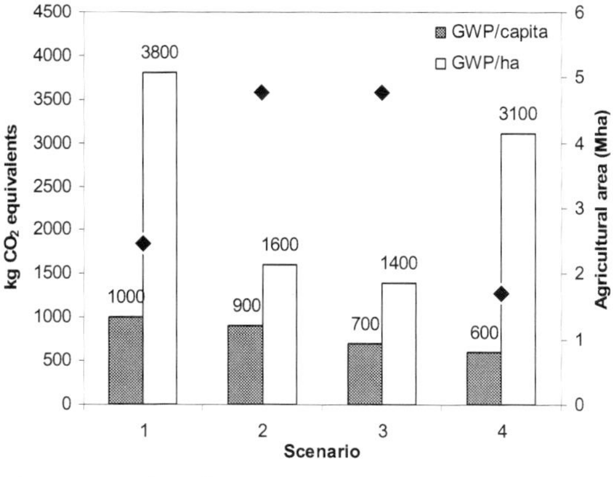

Fig. 2. Global warming potentials in four scenarios, *kg CO_2-equivalents per capita and kg CO_2-equivalents per ha.*

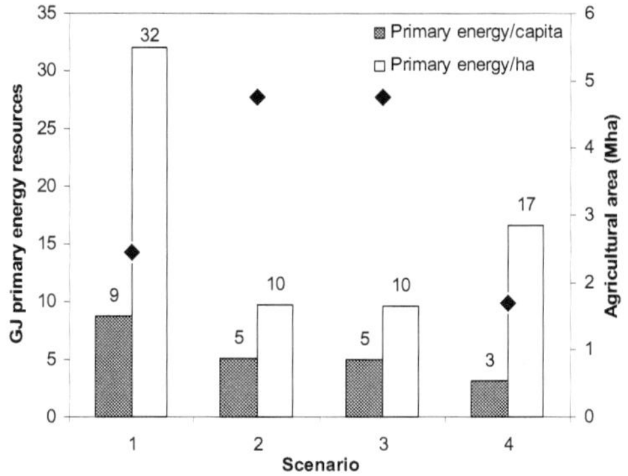

Fig. 3. Consumption of primary energy resources in four scenarios, *GJ primary energy resources per capita and GJ primary energy resources per ha.*

The black diamonds represent the required area for agricultural production, *million hectares.*

4. Discussion

Below environmental and health consequences of different farming regimes and the role of collaboration and consumer demand for sustainable food systems are discussed.

4.1 Environmental aspects of eco-local food systems

According to the Swedish Environmental Advisory Council a diet consisting of two thirds animal products results in four times larger emissions of nitrogen from the agriculture into the water and air compared to a fully vegetarian diet (MVB, 2005). Edman suggests increased shares of local organic food and increased shares of vegetables and a change of meat consumption from monogastric to ruminant meat to reduce the contribution to global warming from the food chain (SOU, 2005). Our study provides results in support of this. The main objective of our consumer survey was to gather data for an environmental impact assessment of an "eco-local" food basket. A food basket consisting of 73% organic food (33% local and organic) and a higher than average proportion of vegetables (100% more) reduced nitrogen surplus with 18 to 37% per capita compared to an average Swedish consumer (Scenario 4 vs. Scenario 1) depending on calculation method used. Thus, not only production methods but also consumption patterns determine the environmental impact.
Simply turning conventional production into a system of ERA without changing consumption patterns would also result in substantial cuts in nutrient emissions. To produce this would however require an additional 2.3 million ha of arable land. This corresponds to a 94% increase and this larger area of arable land is not available in Sweden. Historically the maximum agricultural area in Sweden was about 3.3 million hectares and taking more than this into production again is unlikely. When interpreting the results, it is also important to bear in mind that a large area outside of Sweden is used to produce mainly fodder for the Swedish agriculture. Johansson (2005) finds that 3.74 million hectares are used today for producing food consumed in Sweden. This implies that more than one million hectares are used abroad and that conventional agriculture of today makes use of a larger area than is actually available in Sweden. The ERA-farms are, on the other hand 85 –

100% self supporting with fodder crops. Combining ERA with a more vegetarian food profile (Scenario 4) the acreage needed for food production would decrease from about 2.5 million ha to 1.7 million ha, see Table 3. This opens up for alternative production, e.g. energy, fibre, recreation or export of food products.

What we eat also influences the energy consumed during different stages of the food chain. Generally meat is the most energy demanding food to produce and increased meat consumption is problematic. This is well reflected in a comparison between the different scenarios. Both GWP and consumption of primary energy reduced with a transition towards ERA production (Scenario 2) and with increased vegetable consumption (Scenario 4). If the building up of soil organic matter (Granstedt and Kjellenberg, 2008; Hepperly et al., 2006; Mäder et al., 2002) is considered, green house gas emissions could decrease with 1 500 kg CO_2-equivalents per ha (Granstedt and Kjellenberg, 2011). Following our results some gains were made in terms of GWP by localizing processing and distribution (Scenario 3) but not in terms of primary energy consumption. Pretty et al. (2005) report larger reductions of external effects from localizing production than from switching from conventional to organic production. The referred study was for UK conditions and, in contrast to our study, included a restriction that all food was produced within 20 km of the place of consumption. Other sources, e.g. Sonesson et al. (2010), argue that transportation can be an important contributor to greenhouse gas emissions in the food chain but that the contribution varies a lot. In short, food transports become less efficient the further down the supply chain you get. The last step, consumers' home transports, is the least efficient if cars are used (Sonesson et al., 2010), which often is the case in Sweden (Sonesson et al., 2005). The consumers' transports of food are not included in our study, which could explain the greater importance given to localized production by Pretty et al. (2005). Other potential positive environmental effects of localized production include a reduced need of packaging. Further studies also need to evaluate the reduction of greenhouse gas emissions and other environmental consequences of reduced deforestation in other countries for production of imported fodder and meat products.

By signing the Kyoto protocol Sweden has already agreed to reduce its emissions of CO_2. About 15-20% of the energy consumed is for the transportation of food (SEPA, 1997) and if measures not are taken in agriculture then they have to be taken in other sectors of the economy. There are thus some potential synergy effects of local and organic food production. The relation between distance traveled and emissions of green house gasses is, however, not as clear as one might expect. A study of the Farmer's Market concept (Svenfelt and Carlsson-Kanyama, 2010) shows that, apart from products transported by air, there are no significant difference in energy intensity between food bought at the local Farmer's Market and similar food bought at a supermarket. Although the distance from producer to consumer is much shorter, the transportation to the Farmer's Market is inefficient. Inefficient vehicles are used and there is poor logistics whereas supermarkets are part of an efficient optimized transport system. However, steps could be taken to make transportation more efficient and if the share of locally produced food is increased there is a potential to lower the emissions of CO_2 further through shorter transportation (Carlsson-Kanyama, 1999; Svenfelt and Carlsson-Kanyama, 2010).

Figure 4 presents a summary of the results presented in Figure 1, 2 and 3 showing the relative difference between the environmental impacts in the four scenarios. Scenario 1 (present governance) is set to 1. The dashed bar in Scenario 1 illustrates the 1,3 million ha of agricultural land abroad that Swedish agriculture depends on today.

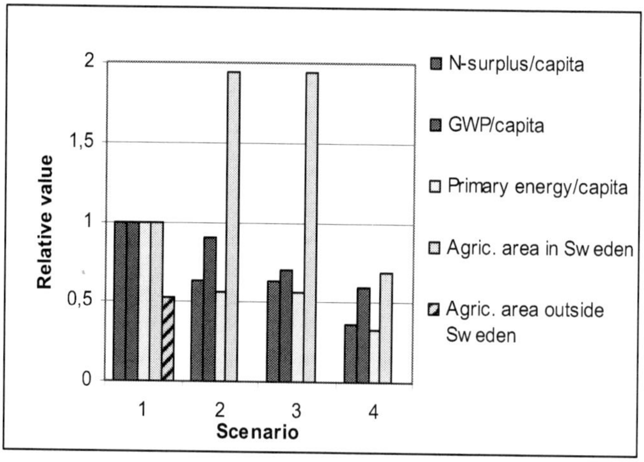

Fig. 4. N-surplus, Global warming potentials and Primary energy resources consumption per capita and required agricultural area in four scenarios, *relative values*.

4.2 Health and sustainable consumption

Our food habits are, unquestionably, important both for our health and for the environment. This is also one of the key issues of Stockholm County Council's S.M.A.R.T. project that gives recommendations for diets that both improve the health as well as decrease environmental impacts (CTN, 2001). New Nordic Nutrition Recommendations (NNR) were approved in August 2004. These are guidelines for the nutritional composition of a healthy diet (NNR, 2004). The NNR do not include instructions for sustainable food choices but such recommendations are available at least in Sweden and in Germany. Some general recommendations include: products produced most nearby when there are equal products; ecological food; less foodstuffs which include few nutrients, e.g. eat fruits instead of sweets (CTN, 2001; SEPA, 1997; 1998; 2000). In Table 4 both nutritional and sustainable food choice recommendations are presented.

The food consumption profile of the studied households seems to follow the diets suggested in the Nordic Nutrition Recommendations (NNR, 2004) and in the S.M.A.R.T. project (CTN, 2001). These households buy a larger share of vegetables (less meat), a larger share of nutritional and storable vegetables (e.g. legumes and root crops) instead of fresh vegetables (e.g. lettuce and cucumbers) during the winter season, less 'empty' calories, more organically produced food and less transported food, compared to the national average food basket. The only large difference is the share of potatoes, see Table 1. The Järna consumers eat substantially less potatoes than the average Swede, while the S.M.A.R.T. project recommends more potatoes. One reason might be recommendations in the anthroposophist nutrient concept - Järna hosts numerous anthroposophist producers and organisations and is considered the anthroposophist capital of the Nordic countries - to minimise intake of solanin producing products like potatoes and tomatoes.

The energy content of consumed (purchased + restaurant meals) 'real' food (excl. sugar, sweets, beverages etc.) was 10.7 MJ/person/day, while the Swedish average 2002 was 10.2

MJ/person/day (Swedish Board of Agriculture, 2004). Thus, we can conclude that our results are in a reasonable range concerning energy content of the purchased food. However, the results are not easily comparable to official statistics due to differences in survey methods.

	Healthy nutrition	**Environmental perspective** **Sustainable food choices**
Fruit, berries and vegetables	- A high and varied consumption of fruit and vegetables is desirable	- A high and varied consumption of domestic vegetables, fruits and berries in season and foodstuffs grown in the field. - If needed off-season, imported fruits or vegetables grown in the field, giving preference to products grown in a nearby country
Legumes		- More leguminous plants instead of meat
Potatoes	- Traditional use, several nutrients, potatoes have a place in a diet	
Cereals	- An increased consumption of wholegrain cereals is desirable	
Fish	- Regular consumption of fish	
Milk and milk products	- Regular consumption of milk and milk products, mainly low fat products are recommended as a part of balanced diet	
Meat	- Consumption of moderate amounts of meat, preferably lean cuts, is recommended as part of a balanced and varied diet	- Less meat - Choose meat from animals that have grazed on natural pasture, e.g. cattle and lamb. - Eat less chicken and pork.
Edible fats	- Soft or fluid vegetable fats, low in saturated and trans fatty acids, should primarily be chosen	- Butter instead of margarine
Energy-dense and sugar-rich foods	- Food rich in fat and/or refined sugars, such as soft drinks, sweets, snacks and sweet bakery products should be decreased	- Eat less
General		- More locally produced food when this is more eco-efficient. - Ecological food - Eat less foodstuffs which include few nutrients, for example: eat fruits instead of sweets - More easily transported foods, e.g. juice as concentrate instead of ready to drink. - Choose the product produced most nearby when there are equal products.

Table 4. Examples of recommendations. (derived from NNR 2004, CTN 2001, SEPA 1997, 1998, and 2000)

4.3 Collaboration, consumer demand and local development

Sustainability in agriculture from an economic perspective requires high quality food at reasonable price to the consumer (Ministry of Agriculture, 2000; SOU, 2004). What defines "reasonable price" is of course a value issue. In Järna higher prices for local organic food is accompanied with high demand. The higher food expense in the consumer survey is somewhat misleading from a societal perspective. The increased cost reflects lower environmental effects compared to conventional food production and consumption where environmental effects to a large degree are externalized. According to Pretty et al. (2005) substantial reductions in external costs could actually be made by a large scale conversion towards local and organic production, similar to the ERA production studied here.

Ecological food is generally more expensive and on a larger scale, higher food prices might hinder a change of consumption. It could be difficult to convince consumers to increase food expenditure for the sake of the environment only and the consumption pattern found in the survey is not expected at most places. A large scale transformation of Swedish agriculture would probably require the government to intervene. This is similar to what is suggested by Edman (SOU, 2004). To increase local, organic, Swedish food production and consumption Edman suggests that the government should strengthen domestic science subjects at school and provide earmarked funding for buying organic food. Out of all food provided by public institutions 25% ought to be organically certified, according to Edman. In Södertälje municipality 14% of the public procurement of food is organic or biodynamic, which places Södertälje among the top five of all Sweden's 290 municipalities (Södertälje municipality, 2006). This share is meant to increase to 50% in 2020. The policy on public procurement from local organic producers is one example of vertical collaboration which facilitates the high concentration of organic farms in the region. It is a good example of a "policy to help nurture green niches and put incumbent regimes under sustainability pressure" (Smith, 2007, p. 447). The collaboration in local, environmentally friendly food systems is not only vertically anchored. Järna community belongs to Södertälje municipality. The families in the household survey are not an isolated group but part of a well developed network of horizontal collaboration. It corresponds well with the local supermarkets having among the highest proportion of sold organic food in Sweden (Larsson, 2007).

The existence of several actors, at various organisational levels, enhances the diversity of governance options (Hahn et al., 2006). In the case of the local ecological food system in Järna different sectors at several levels are involved which could explain why it is so well developed. Quoting Low and Gleeson (1998, p. 189) on environmental governance: "Think and act, globally and locally". Both households and municipalities use their buying power to stimulate local production and development as well as environmental gains through increased demand of local organic food. The consumers' attitudes revealed in the high share of local and organic food and the tolerance towards higher prices could be described as an informal institution based on trust (Svenfelt and Carlsson-Kanyama, 2010) and common norms (Larsson, 2007). The high level of public procurement and the fact that organic farms can lease municipal land at non-market conditions (Larsson et al., 2007) are results of municipal regulations, i.e. formal institutions. These institutions facilitate in a sustainable governance of the community and the local agriculture.

5. Conclusions

We conclude that a sustainable governance of the food system needs to address consumption profiles as well as production methods, since both cause environmental effects. All examined environmental effects were lower on the studied Ecological Recycling Agriculture farms compared to conventional production. Combining this with changes in our food consumption can further reduce the environmental impact of the food system. If all Swedish food production is altered to ERA this would reduce the surplus of nitrogen with 18-37%. In addition to this, if all Swedes were to change their food profiles towards more organic vegetables and less meat the nitrogen surplus could decrease further. Results from our household survey indicate reductions in the range of 55-64% but the number of observations was limited why this should be seen as a special case.

Changing production methods to ERA would reduce emission of CO_2-equivalents and the consumption of primary energy. Combining ERA with an alternative food basket more is won. A change to ERA would decrease the environmental impacts, even when the food consumption profile remains as the Swedish average of today. The agricultural area needed would, however, increase substantially making a large scale conversion less realistic. If coupled with a changed diet the area needed for food production would decrease with 30%.

The results support other findings that changes in food profiles towards a more vegetarian diet and more organic foods decrease the environmental impacts. This change would have a negative effect in terms of increased food expenditures. The families in the household survey consumed substantially more local and organic products, less meat and more vegetables and they spent 24% more money on food compared to average Swedish consumers. Compared to conventional food production and consumption the environmental costs of eco-local food are however to a larger degree internalized.

In the studied community a local food system characterized by a high share of supply and demand of organic food has evolved. This has been facilitated by horizontal and vertical collaboration - horizontal through a high demand from private consumers coupled with large supply from local producers and vertical in the form of public procurement. However, because of the higher price charged for local and organic food a large scale transformation of Swedish agriculture would probably rely more on governmental intervention since few regions experience as high private and public demand for eco-local food.

The environmental benefits of organic agriculture cannot be fully realized unless food profiles change. For a governmental intervention in the form of e.g. public procurement to have optimal effect it is as important to focus on food content as on production methods. Localized processing is however of less importance in terms of environmental effects.

6. References

Adger, N., Hughes, T., Folke, C., & Rockström, J. (2005). Social-Ecological Resilience to Costal Disasters. *Science,* Vol. 309, 12 August 2005. pp. 1036-1039.

Brower, F.M., Godeschalk, F.E., Hellegers, P., & Kelholt, H.J. (1995). *Mineral Balances at Farm Level in European Union,* Onderzoekelseverslag 137, Agricultural Economics Research Institute (LEI-DLO), The Hague.

Cameron, M.E., & Van Staveron, W.A. (1988). *Manual on Methodology for Food Consumption Studies,* Oxford University Press, Oxford.

Carlsson-Kanyama, A. (1999). *Consumption patterns and climate change: consequences of eating and travelling in Sweden,* PhD Thesis, Stockholm University, Stockholm.

Carlsson-Kanyama, A., Sundkvist, Å., & Wallgren, C. (2004). *Lokala livsmedelsmarknader – en fallstudie,* Royal Institute of Technology (KTH), Stockholm.

CTN. (2001). *Eat S.M.A.R.T. – an educational package on food, health and the environment (in Swedish),* Centre for Applied Nutrition, Samhällsmedicin, Stockholms läns landsting, Stockholm. www.sll.se/w_ctn/3938.cs.

Diaz, R.J., & Rosenberg, R. (2008). Spreading Dead Zones and Consequences for Marine Ecosystems, *Science* vol. 321, pp. 926-929.

Dietz, T., Ostrom, E., & Stern, P.C. (2003). The Struggle to Govern the Commons. *Science* Vol. 302:1907-1912.

FAOSTAT. (2004). DOI: http://apps.fao.org/faostat/collections

FiBL Dossier. (2006). *Quality and Safety of organic products. Food systems compared. No.4,* Research Institute of Organic Agriculture (FiBL), Switzerland. http://www.fibl.org

Folke, C., Hahn, T., Olsson, P., & Norberg, J. (2005). Adaptive Governance of Social-Ecological Systems, *Annual Review of Environmental Resources* 30:441-473.

Granstedt, A. (2000). Increasing the efficiency of plant nutrient recycling within the agricultural system as a way of reducing the load to the environment - experience from Sweden and Finland, *Agriculture, Ecosystems & Environment* 1570: 1–17.

Granstedt, A. (2005). Results of plant nutrient balances in the BERAS countries, Concluding results and discussions. In: *Environmental impacts of eco-local food systems – final report from BERAS Work Package 2,* Granstedt, A., Thomsson, O., & Schneider, T. (Eds.), The Swedish University of Agricultural Sciences, Uppsala.

Granstedt, A., Thomsson, O., & Seuri, P. (2004). Effective recycling agriculture around the Baltic Sea. Background report. BERAS 2, *Ekologiskt lantbruk 41,* Swedish University of Agricultural Sciences, Uppsala.

Granstedt, A., Seuri, P., & Thomsson, O. (2008). Ecological recycling agriculture to reduce nutrient pollution to the Baltic Sea. *Journal of Biological Agriculture and Horticulture,* Vol. 26, 279-307.

Granstedt, A., & Kjellenberg, L. (2008). Organic and biodynamic cultivation - a possible way of increasing humus capital, improving soil fertility and providing a significant carbon sink in Nordic conditions. *Proceedings of the Second Scientific Conference of the International Society of Organic Agriculture Research (ISOFAR),* held at the 16th IFOAM Organic World Congress Modena, Italy, June 18-20, 2008.

Granstedt, A., & Kjellenberg, L. (2011). *Skilleby long term trial 1991 – 2011. Final report (in print).* Biodynamic Research institute, Järna.

Gustavsson, A. (2003). Personal communication. Saltå Mill and Bakery, Järna.

Haden, A., & Helmfrid, H. (2004). Järna, Sweden – Community consciousness as the basis for a learning local ecological food system. In Local and organic food and farming around the Baltic Sea. *Ekologiskt Lantbruk* Vol 40, Swedish University of Agricultural Sciences, Uppsala.

Hahn, T., Olsson, P., Folke, C., & Johansson, K. (2006). Trust-building, knowledge generation and organizational innovations: the role of a bridging organization for adaptive co-management of a wetland landscape around Kristianstad, Sweden, *Human Ecology,* 34:573-592.

Halberg, N. (1999). Indicators of resource use and environmental impact for use in a decision guide for Danish livestock farmers. *Agriculture, Ecosystems & Environment,* 30, 17–76.

HELCOM. (2005). The Fourth Baltic Sea Pollution Load Compilation (PLC-4). *Baltic Sea Environment Proceedings,* No 93.

Hepperly, P., Douds, D. Jr., & Seidel, R. (2006). The Rodale Institute Farming Systems Trial 1981 to 2005. In: *Long Term Field Experiments in Organic Farming*, Raupp, Pekrun, Oltmanns & Köpke (eds.), ISOFAR Scientific Series, Berlin.

IPCC. (2001). Radiative Forcing of Climate Change. In: *Climate Change 2001 (6): The Scientific Basis*, pp 349-416. IPCC, Cambridge.

Irish, M. (1982). On the interpretation of budget surveys: purchases and consumption of fats. *Applied Economics* 14, 15-30.

Johansson, S. (2005). *The Swedish Foodprint - An Agroecological Study of Food Consumption.* Doctoral Thesis 2005:56, Acta Universitatis Agriculturae Sueciae, Uppsala.

Larsson, M. (2007). Trust, Adaptability and Community Development - A Case Study of Local Environmental Entrepreneurs. In *13th Annual International Sustainable Development Research Conference. Critical Perspectives on Health, Climate Change and Corporate Responsibility*, Cerin, P., Dobers, P., & Schwartz, B. (eds.), Mälardalen University, Västerås.

Larsson, M., Löf, A., & Hahn, T. (2007). Local organic food system in Järna - ecosystem management and multilevel governance in agricultural production. In: *Governance and Ecosystem Management for the Conservation of Biodiversity*, Manos, B., & Papathanasiou, J. (eds.), Aristotle University, Thessaloniki, pp 101-105.

Larsson, M., & Granstedt, A. (2010). Sustainable governance of the agriculture and the Baltic Sea - Agricultural reforms, food production and curbed eutrophication. *Ecological Economics* 69 (2010) 1943-1951.

Lindfors, L-G., Christiansen, K., Hoffman, L., Virtanen, Y., Juntilla, V., Hanssen, O.J., Rønning, A., Ekvall, T., & Finnveden, G. (1995). *Nordic Guidelines on Life-Cycle Assessment.* Nordic Council of Ministers.

Low, N., & Gleeson, B. (1998). *Justice, Society and Nature: An Exploration of Political Ecology.* Routhledge, New York.

Low, B., Ostrom, E., Simon, C., & Wilson, J. (2003). Redundancy and diversity: Do they influence optimal management? In: *Navigating Social-Ecological Systems: Building Resilience for Complexity and Change,* Berkes, F., Colding, J., & Folke, C. (eds), Cambridge Univ. Press, Cambridge, UK.

Lundgren, L. (1992). *How much is the environment affected by a kWh used in Sweden?* (In Swedish). Vattenfall Research, Stockholm.

Mäder, P., Fliessbach, A., Dubois, D., Gunst, L., Fried, P., & Niggli, U. (2002). Soil Fertility and Biodiversity in Organic Farming. *Science,* Vol. 296 pp. 1592-1597.

Ministry of Agriculture. (2000). *The Environmental and Rural developmental Plan for Sweden 2000-2006.* Ministry of Agriculture, Stockholm.

MVB. (2005). *A Strategy for Ending Eutrophication of Seas and Coasts (Memorandum 2005:1).* The Swedish Environmental Advisory Council/Miljövårdsberedningen, Ministry of Sustainable Development, Stockholm.

Myrbeck Å. (1999). *Nutrient flows and balances in different farming systems - A study of 1300 Swedish farms.* Bulletin from the Division of Soil Management, Department of Soil Sciences, 30, 1-47, Swedish University of Agricultural Sciences, Uppsala.

NNR. (2004). *Nordic Nutrition Recommendations* (4th edition), Nordic Council of Ministers, Copenhagen.

Pena, D., & Ruiz-Castillo, J. (1998). The Estimation of Food Expenditures from Household Budget Data in the Presence of Bulk Purchases. *Journal of Business and Economic Statistics* 16, 292-303.

Pettersson, K., personal communication (2005). Svensk Mjölk (Swedish Milk). Stockholm.

Pretty, J.N., Ball, A.S., Lang, T., & Morison, J.I.L. (2005). Farm costs and food miles: An assessment of the full cost of the UK weekly food basket. *Food Policy*. Vol. 30, pp. 1-20.

SEPA. (1997). *Eating for a better environment. Final report from systems study Food.* (In Swedish.) Rapport 4830, Swedish Environmental Protection Agency, Stockholm.

SEPA. (1998). *A Sustainable Food Supply Chain. A Swedish Case Study.* Rapport 4966, Swedish Environmental Protection Agency, Stockholm.

SEPA. (2000). *Green Purchasing of Foodstuff.* Rapport 5128, Swedish Environmental Protection Agency, Stockholm.

Smith, A. (2007). Translating Sustainabilities between Green Niches and Socio-Technical Regimes. *Technology Analysis & Strategic Management.* Vol. 19, issue 4, p427-450.

Sonesson, U., Antesson, F., Davis, J., & Sjödén, P-O. (2005). Home Transports and Wastage - Environmentally Relevant Household Activities in the Life Cycle of Food. Ambio, vol.34, issue 4-5, pp. 368-372.

Sonesson, U., Davis, J., & Ziegler, F. (2010). Food Production and Emissions of Greenhouse Gases. An overview of the climate impact of different product groups. SIK-Report No 802.

Södertälje municipality. (2006). *Miljöbokslut. Uppföljning av Agenda 21,* Södertälje kommun.

SOU (2004). *Hållbara laster. Konsumtion för en ljusare framtid.* SOU 2004:119, Ministry of Agriculture, Stockholm.

SOU (2005). *Bilen, biffen, bostaden. Hållbara laster - smartare konsumtion.* SOU 2005:51, Fritzes, Stockholm.

Statistics Sweden. (2004). *Household expenditures (HUT) PR 35 SM 0401* (in Swedish). DOI: www.scb.se/statistik/HE/PR0601/2003A01/8%20Hushållsgrupp%20och%20miljö märkta%20och%20ekologiska%20varor%20-andel%20per%20utgiftsgrupp.xls

Statistics Sweden. (2005). *Yearbook of Agricultural Statistics.* Statistics Sweden, Stockholm.

Statistics Sweden. (2011). *Yearbook of Agricultural Statistics.* Statistics Sweden, Stockholm.

Steinshamn, H., Thuen, E., Bleken, Brenøe, M.A., Ekerholt, G., & Cecilie, Y. (2004). Utilization of nitrogen (N) and phosphorus (P) in an organic dairy farming system. *Agriculture, Ecosystems & Environment.* 104, 509–522.

Svenfelt, Å. & Carlsson-Kanyama, A. (2010). Farmers' markets - linking food consumption and the ecology of food production? Local Environment. 15 (5), 453-465.

Swedish Board of Agriculture. (2004). *Consumption of foodstuffs and their nutrient contents. Data up to and including 2002.* (In Swedish). Rapport 2004:7, Jordbruksverket, Jönköping.

Uusitalo, R., Turtula, E., Grönroos, J., Kivisto, J., Mäntylahti, V., Turtula, A., Lemola, R., & Salo, R. (2007). Finnish trends in phosphorus balances and soil test phosphorus. Agricultural and Food Science. 16, 301–316.

Wallgren, C. (2008). Järna Study. Paper 3 in: Food in the Future: energy and transport in the food system. Licentiate Thesis. Royal Institute of Technology, Stockholm.